W9-AHZ-306

ETHICS,
FREE ENTERPRISE,
AND PUBLIC POLICY

ETHICS, FREE ENTERPRISE, & PUBLIC POLICY

Original Essays on Moral Issues in Business

Edited by
RICHARD T. DE GEORGE
AND JOSEPH A. PICHLER
University of Kansas

NEW YORK · OXFORD UNIVERSITY PRESS · 1978

Library of Congress Cataloging in Publication Data

Main entry under title:
Ethics, free enterprise, and public policy.

Bibliography: p.
Includes index.
1. Business ethics—Addresses, essays, lectures.
2. Industry—Social aspects—United States—Addresses, essays, lectures. 3. Social justice—Addresses, essays, lectures. I. De George, Richard T. II. Pichler, Joseph A.
HF5387.E84 174'.4 78-3397
ISBN 0-19-502425-7

Second printing, 1979

Printed in the United States of America

To the memory of
William T. Blackstone

PREFACE

In recent years, ethical issues in business and politics have gained national attention. Political dirty tricks toppled a President. Payoffs to foreign officials brought corporate dismissals. Allegations of professional misconduct resulted in congressional hearings or litigation involving architects, accountants, and lawyers.

Our nation takes morality seriously. To be sure, we have crime and immorality. But these should be seen against a background of general agreement on a large number of ethical issues. Equality and justice before the law are ideals we actively pursue. There is a strong moral consensus against stealing by either the rich or the poor, disregard for the rights of others, fraud, and physical and economic slavery.

The less clear issues generate moral problems. Rapid advances in science, the complexity of economic systems, and the diffusion of responsibility in large organizations have brought forth a host of new moral problems in search of solutions. Is preferential hiring of women and minorities morally justifiable because of past discrimination or is it simply a case of another wrong, two of which do not make a right? Do we have obligations to future generations such that we should use less of our non-renewable resources? If so, how much less should we use and who precisely is included in that "we"? Should public service personnel have the right to strike? Are codes that govern the legal, accounting, journalism, and medical professions really ethical or simply self-serving?

Such micro-moral issues must be addressed within the context of our economic system. Can a competitive system protect the welfare of consumers, managers, and shareholders? Can workers be guaranteed a "fair" wage in an economy that is marked by rapid technological change? How can such a system meet the needs of individuals who are unable to be productive because of illness, age, or limited education? In short, one must also ask the macro-moral question, "Is free enterprise compatible with social justice?"

These issues are the subject of widespread attention. However, behind the newspaper articles and the moralizing of guest columnists, there has been too little serious analysis of such questions. This volume is a step toward filling that gap.

The first two essays are a general introduction to the questions of ethics, business, and public policy. The next three present different but representative approaches to and analyses of the justice of our present economic system. The remaining essays address specific issues of morality and business. Since the issues are complex, no individual essay should be expected to provide a definitive solution. The volume contains contrasting viewpoints, each of which, we believe, helps advance the dialogue necessary to arrive at a defensible moral judgment about the practice or question at issue. Though the theme of morality comes through clearly in all the essays, the sub-theme of the possible role of government in remedying inequities and injustices is noteworthy.

Most of the essays contained herein were originally presented, usually in somewhat different form, at a Symposium on Ethics in Business and the Professions held at the University of Kansas in November, 1976. This Symposium was funded by the University of Kansas; the Kansas Committee for the Humanities; Peat, Marwick, Mitchell and Co.; and the Westport Fund. The remaining essays were written especially for this volume. We are grateful to the sponsors of the Symposium for their initial support. Their interest in promoting wider discussion of important questions of morality in our society has been gratifying. To Karen Allbaugh Snyder, we extend special thanks for her generous assistance.

We hope that others will continue the discussions begun here.

Lawrence, Kansas
May, 1978

RICHARD T. DE GEORGE
JOSEPH A. PICHLER

CONTENTS

ETHICS, FREE ENTERPRISE, AND PUBLIC POLICY

I

CAPITALISM AND MORALITY

1

MORAL ISSUES IN BUSINESS

RICHARD T. DE GEORGE

The position that ethics has nothing to do with business has a long history. But it is a position that takes too narrow a view of both ethics and business. Ethics is concerned with the goods worth seeking in life and with the rules that ought to govern human behavior and social interaction. Business is not just a matter of economic exchange, of money, commodities, and profits; it involves human interactions, is basic to human society, and is intertwined with the political, social, legal, and cultural life of society.

There is an obvious connection between business and ethics. A businessman whose employees rob him blind could no more survive than a businessman who through lying and fraud sold only products that did not work. Such examples, however, do not settle any real issue between business and ethics. For if lying, fraud, or theft lead to the failure of a business, then a businessman might condemn and eschew them not because they are immoral but because they are bad business practices. Actions that are contrary to moral norms may in some, perhaps even in many, instances be bad for business. But we can distinguish acting in a certain way because it is moral from so acting because it is economically profitable. The distinction is between an action done from duty (i.e., because the action is morally correct) and an action which happens to coincide with one's moral duty, done from some other motive. The claim that morality is inapplicable in business affirms that one's calculation should be based on business, not on moral, considerations. Where both coincide, so much the better; where they diverge, so much the worse for morality.

It would be nice to be able to show that moral action is always best for business. But this seems not to be true, especially in the short run. Lying, fraud, deception, and theft sometimes lead to greater profits than their opposites. Hence, moral judgments sometimes differ from business judgments. Businesses need not be run from moral motives. But the actions of businesses affect individuals, society, and the common good. If moral actions are ultimately in the common good, then running businesses in accord with moral norms is in the common good. To the extent that profit maximization conflicts with moral norms, it leads to actions which are not in the common good. In such cases, if the common good is to be protected, and if businessmen do not on their own act in accordance with moral norms, they should be forced to do so. Either public pressure or legal measures should be brought to bear to make the immoral practices unprofitable. If the penalties for not acting morally are sufficiently severe and if the enforcement of norms is sufficiently diligent, then conforming to moral norms will in the long run be in the business interests of business as well as in the general interests of society.

Discussions of morality and business, if the above points are conceded, often lead to three different kinds of questions. First, whose morality is to be imposed on business? Secondly, can economic systems be morally evaluated, and how is morality to serve as a guide between different economic systems and different societies with varying moral norms? Thirdly, which specific practices within a given system are in the public's good?

I

The first question is in several ways a specious one, though seriously and frequently raised. It assumes that there are many different moralities, perhaps that each person has his own, that each is equally valid, and that there is no way to choose among them. When fleshed out as a theory, the position has come to be known as moral relativism. As a philosophical position, it has been refuted at some length,[1] and I think cogently. For present purposes, however, we need look only at the argument based on the fact of cultural diversity and on the fact of moral pluralism within our own contemporary American society. Do either of these facts provide adequate grounds for arguing that there is no morality we can collectively and appropriately apply in the moral evaluation of business practices? I think not, for three reasons.

First, despite the differences in many customs and moral practices from society to society and age to age, there is and has been basic agreement on a large number of central issues. The most basic agreement, though completely formal, is that good should be done and evil avoided. Substantively, there are moral requirements that the members of any society must follow if it is to survive. These include, among others, respect for the lives of its members; respect for the truth (without which it would be impossible to communicate and carry on social life); and respect for cooperation and helpfulness. Each society, for instance, holds that it is immoral for any ordinary member without good reason to kill other members of that society. A society may not have any compunction about killing members of other societies. But unless a society holds that killing its own members without sufficient reason is wrong, the society could not long endure.

Secondly, even in those cases in which the members of a society are allowably killed—e.g., executions, infanticide, euthanasia—the society usually has a network of beliefs which justifies the practice. Beliefs differ, however, and some people believe what is false. A false belief may lead to a different practice than a true belief, but mistaken beliefs do not validly ground a practice. The fact of differing practices, therefore, does not by itself support the claim that there is no way to decide between them. Sometimes there is. In those cases in which moral practices are based on false beliefs, bad reasoning, inconsistent premises, and so on, we have a means of sorting out practices and choosing or rejecting one rather than another. Progress in morality and changes in moral practices, in fact, frequently follow upon the growth of knowledge and changes in belief. The basic similarity of human beings, for instance, is now sufficiently clear and established that we can rightly say it is morally improper to exclude women, or blacks, or men, or any sub-group from the moral community. Criteria of sex, race, nationality, and so on are not morally relevant criteria. This was true even before it was recognized. Unless it was true, it is difficult to understand how it could have come to be recognized.

Thirdly, many differences in moral practices from culture to culture can be shown to be the result of differing circumstances, not of differing moral principles. Morris Ginsberg argues this persuasively and at length.[2]

Thus, the fact that different parts of the world have somewhat different moral practices shows neither that they are all equally valid nor that any one of them is necessarily invalid or immoral.

What about moral pluralism within our own society? Does it justify the claim that all moral judgments are equally good or valid, sound or proper? No, it does not. Reasoning analogous to that used above with respect to differences between cultures applies here as well. For differences are data to be explained, not premises from which moral relativism can be validly deduced.

In American society, despite the rather large amount of disagreement on some issues—e.g., abortion, birth control, capital punishment—there is an enormous amount of agreement. This is true of any viable society; for without such agreement the fabric of society would disintegrate. Not only is there general agreement that murder, theft, lying, fraud, and so on are immoral, but the public reaction both to the events connected with what has become known as "Watergate" and the public reaction to bribes by corporations is evidence of a large degree of agreement on moral issues such as these. That peoples in some other parts of the world did not understand American public reaction in these cases might indicate that they are cynical about morality, that they do not expect national leaders or businessmen to behave morally, or that they hold different moral values. But their reactions are beside the present point.

In those realms where there is real difference of moral opinion in our country, moreover, there is some consensus about how to deal with that difference. Thus, there is general agreement that morally different opinions are to be tolerated and actions not legally controlled if their results fall exclusively or almost exclusively on the consenting, informed adults who participate in the actions. Fornication and adultery are two cases in point. Other actions, though perhaps not considered immoral, may harm persons other than the participants. Here, there is general agreement that the use of laws to prevent such harm is appropriate. Whether a particular law gets passed, however, depends on the support which can be generated for it through debate, argument, and other means (e.g., lobbying). The recent history of abortion laws is an example of how differing moral judgments are handled where there is disagreement about whether the rights of innocent persons are being infringed.[3]

The pertinent conclusion for our purposes is that the morality which is to be applied to business in our society is the morality which is generally held by the members of our society; that this covers a large area on which there is agreement; and that, on those topics on which there is disagreement, there should be informed debate about the nature of the activity in question, the circumstances, and the moral principles which differing groups think are applicable. Progress is made this way in other areas, and it is the only way that progress can be made in this area as well. Whether slavery is immoral is not just a matter of opinion; whether discrimination is immoral is not just a matter of opinion; whether bribery is immoral is not just a matter of opinion. These are all matters on which a consensus has been reached (at least in our country, though also in a great many other parts of the world as well), and reached for good and substantial reasons. That these and other moral judgments cannot or should not be applied to the realm of business is simply not true.

The morality held by the American people is expressed in their moral beliefs, institutions, and laws. In areas where there is no specific legislation, the norms implied in legislation should be applied. In disputed areas, the moral judgments which can be best defended and articulated should hold sway. Yet, three points should be made. First, morality is broader than legality. Not everything that is immoral can or should be made illegal. Public reaction, however, is not expressed only in law. Informed public opinion can affect business, and improper business activities should be protested in the market as well as in the ballot box or in the legislature.

Secondly, members of the business community are also members of society. Both should and usually do have a sense of what is immoral, of what they would perceive to be immoral in the business activities of others, and of what the public would perceive as immoral in their own business activities. Just as individuals do not exist in a vacuum but in a society, so businesses do not exist in a vacuum but in a society. As part of that society, they, as individuals, have responsibilities to the society as well as to themselves. Ignoring unethical practices when engaged in by others so that they will be ignored when engaged in by oneself leads to the undermining of confidence in business by the public and will eventually lead to legislation or the wider public enforcement of moral norms.

Thirdly, the view that business practices should be restricted not by moral considerations but only by legal considerations fails to recognize two points. First, one of the bases for making practices illegal is that they are immoral. Second, if immoral practices are not policed by an industry but are engaged in until made illegal, the resulting legislation is frequently excessively restrictive, over-determined, and more costly than uncoerced compliance with moral norms. Thus, industry-wide self-regulation is frequently preferable to legislation. Self-imposed codes of ethics, if honestly drawn up and followed industry-wide (because policed by the industry)—providing they are not self-serving or a means of controlling that legitimate competition which benefits society—can both protect the public interest and keep the conditions of competition within an industry fair.

Our discussion thus far has not addressed itself to borderline cases, and has not suggested that it is always clear exactly what is moral and what is not. It has argued in various ways, however, that despite these uncertainties there is a large area of agreement about the morality of business practices and that the claim of ignorance about what is moral, based on moral pluralism either cross-culturally or within our own society, is largely spurious.

II

Following the economic model, the second group of questions arising from the connection between morality and business can be called macro-moral. These questions involve the moral evaluation of entire economic systems or the moral evaluation of countries and peoples vis-à-vis one another in the economic realm. Issues of the second type arise, for instance, from the fact that there are at the present time some rich countries, some poor countries, and some countries in between. Wealth may come from natural resources or from industrial development. There are some persons who question the right of any people or of any nation to control the resources of a given territory when these resources are needed by others, as well as the right of a manufacturing country, for instance, to exploit the natural resources of another country. An example of the first type would be whether oil-rich countries are morally justified in collectively raising the price of oil to any level they like, regardless of the effect on the oil-using nations—both the relatively well-to-do ones and the poor ones. Is this a moral question at all, or only an economic one? If human welfare is affected by these actions, as it obviously is, then there does seem to be a moral dimension. But exactly how moral suasion can be brought to bear among nations and peoples is far less clear than how it can be brought to bear in a given society. From a moral point of view, the problem is exacerbated by the fact that we are not dealing with individuals but with countries and peoples; how responsibility is to be assigned is consequently less clear than when dealing with individual cases.

The matter of economic exploitation poses a similar problem. Suppose that a manufacturing country did at some time in the past exploit the raw materials of a colony or of an underdeveloped country. Does the former country now owe reparations to the latter? If so, who precisely owes what, if the exploitation is no longer practiced and those who practiced it are now dead? How exploitation is to be defined and how one is to decide whether it was in fact or is in fact practiced remain problematic. But exploitation aside, whether one country which enjoys a high standard of living has any responsibility to help another country which has a low standard of living or whose people are unable to subsist on what they have is a large moral problem, the dimensions of which have yet to be adequately described.

I shall not attempt to solve these problems. Rather, I shall briefly turn to a second type of macro-moral problem, viz., whether economic systems can be morally evaluated.

Hardly anyone would deny that we all have a *prima facie* moral obligation not to engage in immoral practices. So if a given economic system is built on an immoral practice or practices, we have a *prima facie* moral obligation not to engage in the system. But showing that any economic system—which is always bound up with the legal, political, and social system of a society—is both inherently immoral and practicably replaceable is frequently no easy feat.

We can rest secure in the moral judgment that slavery is always *prima facie* morally wrong and to be avoided. We can be less sure that because they were built on slavery, ancient Greece or Rome or Egypt were immoral societies. For we may have some doubts about whether, in those cases, there were any viable better alternatives. I am not arguing that there were not; I am simply affirming that not every *prima facie* immoral practice must be avoided, since some such practices may be the least bad of the available alternatives. Nor should the judgment that slavery is immoral be confused with the judgment that everyone in a slave-holding society is acting in a subjectively immoral way. For they may not realize that the practice they are engaged in is immoral. Yet despite these caveats, we can be sure that slavery is an immoral economic—as well as social and political—system for present-day Americans to adopt.

Granted, therefore, that economic systems can be morally evaluated, we can legitimately ask: is the capitalistic system inherently immoral? Though it is not clear that Marx himself raised moral objections to capitalism, Marxists certainly have, and some of them have come to a negative conclusion. How valid is this assessment?

Four arguments have been used to show the inherent immorality of the capitalist system. One holds that capitalism is built on the exploitation, and so in a sense on the cheating or robbing, of either its own workers—the proletariat—or those of less developed countries. A second and related argument says that capitalism involves a kind of slavery—wage slavery—which is as immoral as physical slavery. A third claims that capitalism yields alienation and similar ills and is therefore immoral. The fourth claims that though capitalism may not have been immoral when it was necessary to develop the productive resources which alone could raise the standard of living of mankind and make a better life possible, it is now restrictive, hampers further development, could and so should be replaced by a preferable—from a human and so moral point of view—alternative. The last claim is not that capitalism is necessarily immoral but that it is immoral to choose it over some better available alternative.

We can examine each claim in turn.

The claim that capitalism is built on the exploitation of workers is sometimes attributed to Karl Marx. Marx's interpretation of the labor theory of value holds that the only way an entrepreneur can make a profit is by paying his workers less than they produce. He does not claim this is immoral. For they are paid what their replacement value is, and so in that sense they are paid what they are worth. But the entrepreneur profits by the discrepancy between what the workers produce and what he pays them, and, if successful, he, not they, gets rich.

The view that maintains that the appropriation of surplus value by the entrepreneur is stealing and so inherently immoral, however, must hold that all value belongs by right to the person who produces it. There are a number of good reasons why this view should not be held. First of all, it does not take sufficient account of how machines multiply the value produced by human labor. An individual worker who has to gather his own raw materials, fashion his own tools, and make whatever he produces individually should get the value of what he produces. With industrialization, however, the labor power of a worker is multiplied by the machines he uses. If a worker today using machines can make ten times as many shoes as he could make without machinery, does it follow that he should be paid ten times as much? It would seem not. For in the first place, as Marx saw, someone has to pay for the machines. More importantly, it took the creative genius of someone—not the particular worker who uses the machines—to invent the machine. It also took the initiative of the entrepreneur who risked his capital and the invested capital of the stockholders who put up the money for the enterprise. So although the worker is able to produce a great deal of value in a short amount of time due to the multiplication made possible by machines, all the credit should not go to him, nor therefore should all the value of what he produces. On this first interpretation, if all of the owner's profit is stolen from the workers, he gets nothing for initiative, risk, inventiveness, and so on.

A second interpretation maintains that the owner's profit is not entirely theft but is so when excessive. With the rise of labor unions, workers banded together to force management to share some of the profits that an enterprise realizes, to improve working conditions, to shorten working hours and days, and to get legislation passed that favors or protects them in various ways. If the difference between what the owners and managers of an industry get and what the lowest paid worker gets is too great, this may be unfair for some reason. A system of taxation or some other form of redistribution may well be in order. But this is compatible with private ownership of industries and admits that a certain portion of surplus value may legitimately go to the owners of industry.

Nor for similar reasons can the case be made out that capitalism or some variety of it has been able to continue because it has shifted its exploitation from its own workers to underdeveloped countries and it is those workers who are being exploited. The argument concerning individual workers in the country in question parallels the above analysis. If it is claimed that the country as a whole is exploited by the richer country, then the technique of exploitation must be characterized otherwise. If this was the case in the relation between mother countries and their colonies, then it does not seem to have been the case *because* of capitalism. Industrial countries today still import raw materials from other countries which are sometimes less developed. This fact by itself does not equal exploitation. Moreover, the raw materials can be sold to countries run more or less on the capitalistic model or to countries run on the socialistic model. The price at which the materials is sold does not depend on the economic system of the buyer country. Hence, the case cannot be sustained that the capitalistic countries depend on exploitation unless the socialistic countries are to be blamed likewise. The evil, in that case, would not be an evil of capitalism, but an evil of well-to-do or industrial as opposed to poor and non-industrial countries. This kind of charge would then become the kind we looked at above; but it would not be the kind of claim we are presently examining, viz., that capitalism as a system is inherently immoral. The claim that exploitation is necessarily ingredient in capitalism and that it is inherently immoral, therefore, does not hold up well under analysis.

The second attack—namely, that capitalism involves wage slavery and that wage slavery is almost as bad as absolute physical slavery—is also defective. The view might be plausible if it could be shown that capitalism pushes workers down to a lower and lower level of life; that they have no bargaining power and are dependent for their subsistence on selling themselves to the owners of the means of production; that they work under inhuman conditions; that they are forced by the system to work at whatever job is available; and that not only heads of households but all members of a household, small children included, are forced to work under inhuman conditions. This is the situation Marx so poignantly describes in *Capital.* Had his analysis that the plight of the workers would get worse and worse until they had nothing to lose but their chains been true, then it could well be argued that the workers had a moral right to seize the instruments of production and change the system. For the system would not allow them to develop their human potentiality, would deprive them of human dignity, and would rend society dangerously apart. If that is the result of capitalism, then it should be morally condemned. Everyone in the society would be alienated, set one against another, lose his sense of human dignity, and be alienated not only in labor but also in society, in politics, and so on. But once again, that situation does not seem to be a necessary ingredient of what has come to be called capitalism. If there ever was such a situation as that described (it never got quite that bad, and so Marx's revolution never took place in the capitalist countries of the West), then it could be branded immoral, just as slavery is branded immoral. But that situation does not prevail today, and we cannot therefore validly make the claim that the economic system of the so-called capitalist countries is inherently immoral because it involves workers in such dehumanizing, slave-like conditions.

The third claim is that private ownership of the means of production results in the alienation of man from the product of his labor, from the labor process, from himself and other men. There is something wrong with a society that values goods more than people, that dehumanizes people in the labor process, and that fragments human beings into competitors, preventing them from social cooperation and mutual respect. And if it could be shown that these were the results of private ownership of the means of production, we would all have to agree that any system built on such ownership is inherently immoral. The link between such conditions and their source, however, is not a conceptual one but an empirical one. And it seems clear that in those societies in which the private ownership of the means of production has been done away with, there is no perceptible decline in alienation, in the desire for goods, in the dehumanization that is tolerated in factories, and so on. And in societies which continue to have private ownership of the means of production, we find growing numbers less interested in goods than their parents; we find a stronger defense of human rights than elsewhere; we find a consciousness that certain types of work can be dehumanizing and stultifying and attempts to change such conditions; and we find not only competition but also cooperation and a willingness to work together.

How we deal with alienation and dehumanization are real problems which we must face. But that they are the result of private ownership of the means of production and will disappear with the disappearance of such ownership seems clearly not to be true.

The fourth charge has the most plausibility. The claim is that capitalism was not (and so is not) inherently immoral. It was historically necessary to develop the industrialization without which mankind could not produce the wealth and the goods necessary for it to increase its standard of living and of dignity. But the capitalist system now serves as a brake, precluding the distribution of the wealth it produces for the benefit of all mankind; it fosters conspicuous consumption; it continues alienation when it would be possible to eliminate or at least substantially reduce it; it protects the vested interests of the rich at the expense of the still poor; it stands in the way of change. It therefore prevents the development of a better, fairer, juster, more moral, humanly preferable society. The charge is not that it is inherently unjust but that it is morally wrong because it inhibits a system which is morally preferable.

Several claims are bound up in this charge, and it is well to separate them out. The first is that there is a better system which would be non-capitalist, or at least non-capitalist in the sense that capitalism exists in some countries today. What the more moral alternative would be and how to achieve it, however, are crucial questions. The Marxist position is that the more moral system would be communism, which would have as its first essential ingredient the abolition of private ownership of the means of production. But what is to take its place? The model of the Soviet Union suggests that state capitalism takes its place. Are the workers better off in the Soviet Union than in the United States? No. Do they enjoy a higher standard of living? No. Do they have more control over their work and working conditions? No. Do they have a less repressive government as a result of their non-capitalistic economic conditions? No. They have, in some respects, more security. But that they have more freedom for self-development and self-expression can at least be argued. Alienation has not disappeared with the abolition of private ownership of the means of production. So if the Soviet model is what is to replace the capitalist model, not only from a human point of view but even from an economic point of view, the case has not been made that capitalism as practiced in the United States is preventing Americans from enjoying the better life available to them if they switch to the Soviet model. The planned economy of the state-owned system has not shown itself superior to the individual initiative still possible in less planned and at least partially privately owned systems.

The Soviet model, of course, is not the only possible alternative, though it is one frequently suggested by Marxist critics. A system of non-Soviet socialism is another alternative; a mixed system is a third possibility; and there are others. But among these, we should consider each on its own merits, first to see if it is in some ways superior. We have no clear model that has developed historically, no particular society which we can turn to as the moral model to emulate.

Secondly, if we did have a model, we would still have the question of how to arrive at it in fact. Seizure of the means of production in the United States, for instance, seems both unlikely and implausible. For what would the workers do once they seized control? Workers, through their retirement plans, presently control large portions of the stock of major industries. What they want is what every stockholder wants—they want their stock to go up. They are not demanding that the companies be run differently. The counter might be that they must be taught to want the companies to be run differently. If there is good reason for such a claim, then they should be taught. But exactly what that way is, is still not clear. A workers' revolution is not in the offing; and a led revolution runs all the dangers of ending up with a totalitarian government at the helm.

The American system is clearly not without fault; it has immoral aspects, it includes immoral practices, it does not achieve distributive or social justice, it is not without corruption, it contains unjustifiable inequalities of wealth, treatment, and so on. As these are slowly or more vigorously corrected, we may develop a kind of society that no longer deserves the name of capitalism. But the arguments we have looked at do not show that the system as a whole deserves moral censure, nor do they present us with a better viable alternative. We have no moral panacea to apply; we have no utopia waiting to be grasped or formed. But we do have possibilities for improvement which will make our society, and so our system, more moral. Piecemeal change, however, from a moral point of view, brings us away from what I have termed the macro-moral questions of evaluating worldwide problems of justice and the morality of systems, to a consideration of micro-moral problems—e.g., the morality of particular practices, states of affairs, and so on within a given system.

III

Micro-moral problems are not less important than macro-moral problems and do not deal necessarily with small issues, but they take as their frame of reference the existing situation and seek to make judgments within the system rather than about the system.

Since I cannot in the space of a short essay even begin to enumerate the vast array of micro-moral problems, much less solve them, I shall simply categorize a few of them.

The problems of morality in business, taken within a given framework or economic system, can be divided broadly and somewhat arbitrarily into six kinds.

The first of these concerns the determination of the justice of distribution of resources. It involves questions of the right to property, questions of ownership and use, questions of just wages, return on capital investment, appropriate reward for risk and skill and inventiveness, as well as questions of providing for the members of a society so that they do not fall below a certain level of welfare or respect. All these questions demand for their solution prior agreement on what in general constitutes justice for the society, as well as how justice is to be weighed when it comes into conflict with welfare, liberty, and other values which are socially important. The development of a theory of justice, not only an ideal theory but also a theory by which we can determine present-day injustices and the proper remedies for past injustices, is required. We do not have an adequate comprehensive theory as yet. Many competing claims are made even in the name of distributive justice: allocation according to equality, need, effort, achievement, contribution, ability, and so on. These must be weighed and balanced as best we can with each other and with commutative and other kinds of justice.

Even with a definite notion of what is just, a second set of questions arises in trying to apply any given concept of justice to particular cases. Since no two cases are exactly alike, it is often difficult to know exactly how to apply the principles. Easy cases can be handled. The more difficult ones pose moral problems which can be resolved only by discussion, debate, and ultimately by making some decision on the basis of the best information and insight available.

A third group of moral problems arises from a conflict among different values, especially when there seems to be no good alternative. Justice, security, liberty, productivity, efficiency, and other values frequently conflict, with no optimal solution available. Here the determination and choice of the least bad alternative is the best one can do.

The fourth kind of moral problem comes from the development of moral insights and the task of applying them to previously accepted practices. For a long while in the United States, for instance, segregation was generally accepted, at least in some parts of the country.

Though now seen as immoral, and though it is now recognized that it was immoral when practiced in the past, it took a long while and a whole program of reeducation to get people to see this. Discrimination and sexism are other cases of this kind. How many other practices do we engage in without the consciousness of their immorality? Some claim that our treatment of animals, especially in slaughterhouses, is immoral. Only in the fairly recent past have we come to consider the industrial pollution of the environment to any serious degree immoral, and have we acted to make some instances of it illegal. It is not always clear who is morally responsible. If the pollution of the air by automobiles is injurious to the health of all of us, do we have a moral obligation to drive less or not at all, or to force manufacturers to build cars with antipollution devices? Are the manufacturers morally culpable if they produce automobiles which do not have such devices?

The fifth general kind of moral problem arises from new activities, products, and techniques. The problems raised by scientific advances are in some ways new problems. We now have the capability of wiping out the human race, of despoiling the environment in a way that will make it unfit for future generations, of using up unrenewable resources, of manipulating genes, and so on. Each of these problems has aspects that affect business: whether industries should engage in the production of certain materials and whether workers should work in such industries.

Lastly, and perhaps most prosaically, are the questions of the application of ordinary accepted moral values and prohibitions in the conduct of business and industry. It is generally accepted that lying is wrong; that stealing is wrong; that harming another is wrong; that giving and taking bribes is wrong; and so on. Each of these leads to specific cases in business as well as in other branches of life. If lying in advertising can make more money for a company, should a company do so? What exactly constitutes lying in advertising? How near to lying can one come? How much exaggeration for emphasis is allowed? How much must one disclose about the bad parts of one's products? Is telling half the truth lying? Whether these questions are asked about a corporation or about members within a corporation—either of management or of workers—they are not new questions; nor do they raise really new problems. The obligation to be moral applies to all aspects of our lives, and our business activities are not exempted. This does not mean that morality is always observed. And where it is not self-imposed, there is the need for its being imposed from the outside with appropriate sanctions to preserve the common good.

The case of bribery to foreign government officials in the recent past is interesting. The argument advanced in defense of the practice was that in dealing with certain foreign officials, since bribery was a way of life for them, a business could not successfully compete unless it paid bribes; if it did not, other companies from one's own nation or from other nations would pay them and win the contracts.

When the practice was disclosed publicly, it raised a small furor in the United States. It led to the dismissal of a number of officers of corporations. Some claimed that it would place American businesses at a disadvantage. The results, however, are noteworthy. Gulf Oil was involved in such practices, and several members of the Board were forced to resign. Since then, the corporation has been following a strict set of self-imposed rules with regard to bribes. Despite this, it has been able to compete successfully. And in some countries in which bribes were previously the way of life, officials have come under pressure themselves and have been forced to reform the practice. Public disclosure of immoral practices in this instance served to help reform the practices not only of American companies but of companies and governments in other parts of the world as well.

IV

Moral philosophers do not have all the answers to all moral problems and dilemmas. Some moral philosophers are skilled at moral reasoning; some know the ways by which moral claims can be defended or presented better than those untrained. But they do not have any monopoly on knowledge or moral values. They cannot and should not be expected to solve moral riddles. But they can be expected to clarify moral alternatives, articulate moral values, and pass on techniques of moral reasoning.

If my claim is correct—that the morality which should be applied to questions of business ethics is the socially accepted morality, as open to correction and development—then disputed questions should be discussed openly and publicly. The public debate should be articulate, informed, intelligible, clear, and should proceed in a rational manner if a rational conclusion is to be reached. It should be free of demagoguery, or at least, people must learn how to distinguish the demagoguery from the reasoned argument. People can and should be trained to think clearly on moral issues, just as in other areas. The moral philosopher can attempt to develop a theory of morals which accounts for and coheres with our consistent moral intuitions, and there are disputes about which theory does this best. But most theories agree in large part on which actions are moral and which are immoral. We cannot wait until all the theoretical issues are all solved before we undertake practical moral problems. Moral education helps all the members of a society take part in the debate about the common good, about the values to be realized and sacrificed, about the balance to be struck between justice and welfare and liberty, and so on. It also helps each member of society to be more conscious of his own responsibility to act morally; it motivates him not to engage in immoral practices either on or off the job; and it makes him sensitive to the possible immoral practices that can be found in the business world as well as in any other sector of life.

Many facets of our system have not yet caught up with our social needs, and the people as a whole, as well as the managers of business, have not yet faced up to the social responsibilities—much less devised techniques for handling them—of business. It is inherently easier to decide how to produce a product at the best price than to worry also about the social impact of that decision. But the wags who wink at immorality in business, arguing that ethics and business are two separate spheres and never the twain shall meet, are short-sighted and look only as far as the last line of their financial statements. Society is larger than that; business is part of society; and ethics has as much a place in business as in any other part of social life. When all members of society realize that fact and act accordingly, society will be that much better off.

NOTES

1. See, for instance, W. T. Stace, *The Concept of Morals,* N.Y.: The Macmillan Company, 1937; Paul Taylor, *Principles of Ethics,* Encino and Belmont, Calif.: Dickenson Publishing Co., Inc., 1975; Brand Blanshard, *Reason and Goodness,* London: George Allen & Unwin Ltd., 1961. A useful collection of essays on the topic is John Ladd (ed.), *Ethical Relativism,* Belmont, Calif.: Wadsworth Publishing Co., Inc., 1973.
2. See his *Essays in Sociology and Social Philosophy,* vol. I: *On the Diversity of Morals,* London: William Heinemann, Ltd., 1956.
3. For further development of this point, see my article, "Legal Enforcement, Moral Pluralism and Abortion," *Philosophy and Law: Proceedings of the American Catholic Philosophical Association,* XLIX (1975), pp. 171–180.

2

CAPITALISM IN AMERICA:

Moral Issues and Public Policy

JOSEPH A. PICHLER

THE ECONOMIC PROBLEM

No society can remain stable for any appreciable period of time unless it solves the "economic problem." This problem has two parts. First, human and material resources must be allocated toward the production of goods and services in quantities sufficient to satisfy at least the minimum expectations of society's members. Second, the resulting output must be distributed among the population.

A wide variety of economic models are available to perform these material functions. In choosing among them, a society is generally guided, in part, by moral considerations. Both the allocation and distribution systems must be consistent with society's shared norms of justice and personal freedom. Both must be congruent with the political structure that reflects those values. If an economy fails to meet its material goals, society will be poverty-stricken. If it fails to satisfy the moral criteria, significant segments of the population will consider the system to be unfair. In either case, the result is misery and social instability.

This essay considers the capitalist form of economic structure, both as an abstract system and as it has been implemented in America. The first section defines capitalism's preconditions and analyzes its dynamics. The second reviews the Constitutional foundations of free enterprise.[1] The third and fourth sections, respectively, discuss the expression of ethical values within a capitalist system and the manner in which American public policy has guided the economy toward the achievement of certain values. The conclusion summarizes the relationship between capitalism and morality.

CAPITALISM AS A SOLUTION: Preconditions and Material Implications

Free enterprise is a system that attempts to solve the economic problem by integrating the allocation and distribution systems through market prices. Economic units have an incentive to acquire or produce those resources and goods that bring the highest price relative to cost.[2] The price received by the seller is income which, upon expenditure, distributes the goods and services produced elsewhere in the economy. Thus, each market transaction simultaneously allocates resources toward productive ends and generates income that will distribute society's resources and goods.

The structural preconditions necessary to the existence of capitalism are private property, contract enforceability, a recognized medium of exchange, and a competitive market structure. The last of these, competition, is defined in terms of four additional conditions: prices (including wages) are free to rise and fall without interference from government or private associations; resources and products are free to move among industries and geographic locations; resource owners and producers of goods have reasonably complete information on transactions and prices; and each economic unit provides such a small relative share of a particular good that it cannot affect the market price.

If these preconditions could be satisfied fully, resources would be allocated and products distributed in accordance with consumer choice. In such a system, individuals are free to purchase goods which, in their judgment, offer the greatest expected satisfaction relative to price. Their choices cause the price of any good to rise or fall in accordance with the relative degree of satisfaction it provides. At the same time, economic units utilize resources to supply goods that yield the highest price relative to cost. In this way, resources are allocated toward goods that satisfy consumers to the greatest degree. As consumer tastes change or new products become available to satisfy existing needs, price/satisfaction ratios are altered and resources reallocated. Thus, the incomes of resource owners and producers of goods fluctuate in accordance with their success in meeting consumer preferences.

The importance of each precondition may be shown by considering the effect of its absence. If property cannot be held privately, resource allocation decisions must be made by a group process rather than by individual choice. Some form of incentive other than profits must be instituted because economic units are not allowed to retain the proceeds of favorable exchanges. Market transactions do not simultaneously allocate resources and distribute output under these circumstances.

The absence of contract enforceability and a recognized medium of exchange introduces so much uncertainty and makes the cost of transactions so high that the price system is inoperable. If contracts are unenforceable, then negotiations for future exchanges are pointless and all transactions must take place "on the spot." There is no penalty for fraud other than private retaliation. In such a world, economic units would hesitate to allocate resources toward the production of specialized goods because the price could not be negotiated in advance. Similar problems are introduced if there is no recognized medium of exchange. The price system collapses into a barter economy wherein each transaction is resource or product specific on both sides of the exchange. Generalized prices cannot be known and, hence, cannot guide allocation decision.

Consumer choice also fails to guide the allocation of resources and the distribution of wealth if any of the four elements of competition is absent. Prices that are limited by government *fiat* or cartel action cannot fully reflect consumer preferences.[3] If the price of a good is set above the market level, economic units are induced to provide excessive quantities that will be wasted or sold on the black market below the fixed price. If prices are set below the market rate, economic units do not receive sufficient incentive to provide all of the goods and services that are demanded. In this case, black market sales will be made at prices above the fixed rate.

Gluts and shortages, relative to consumer desires, also develop on a local or societal basis if resource owners and producers do not have accurate information on transaction prices or if there are legal barriers to movement among industries and geographic locations. Differential prices for the same good indicate differential preferences or resource costs. The profit incentive can induce economic units to arbitrage these differences only if they are known and the unit encounters no barriers other than transportation costs to moving among exchange partners.[4] Finally, it may readily be shown that an economic unit large enough to influence the price of a good, either by itself or in combination with others, will find it profitable to provide less of it than the amount that would allow consumers to reach their desired price/satisfaction ratio.[5]

Capitalism also makes certain assumptions about human motivation. The most important are that individuals seek to maximize their own self-interest, that the satisfaction of material needs is a significant element in the definition of self-interest, that few members of society have sated their material desires, and that not every good is perceived to be equally satisfying. In the absence of such conditions, there are no consumer preferences to drive the economic system. If the assumptions are valid, then the implementation of capitalism's structural preconditions provides a powerful, but noncoercive, incentive for economic units to allocate resources in a manner that satisfies consumer desires most efficiently. This statement deserves some elaboration.

First, the process is purely voluntary. Capitalism operates by incentives that "draw" resources toward the production of goods preferred by consumers. As economic units freely allocate resources toward profit opportunities, they serve consumer interests as well as their own.

Second, consumer preferences are the sole determinant of price and profit incentives. At first sight, this statement might appear to be incomplete because the price of a good, and the profit earned from its production, must surely be affected by resource costs. Because economic units have an incentive to supply a good only if price exceeds cost, it would seem that market price and profit are a function of resource costs independent of consumer preferences. However, careful analysis will show that, for any given state of technology, resource costs are *themselves* determined by consumer preferences. The cost of any resource used in the production of a good is equal to the highest price that economic units will pay to use that same resource in the production of some alternate good. But the price of the alternative good is itself determined by the price consumers are willing to pay for it. Thus, consumer preferences determine the cost of resources, and it is strictly correct to say that, under capitalism, consumer preferences determine prices and profits.

Third, capitalism provides an incentive for economic units to satisfy consumer needs in the most efficient manner. Consumers find it in their own interest to purchase a desired good of given quality from the economic unit that makes it available at the lowest price. The lowest price can be charged by those who produce at the lowest cost. As a result, economic units have an incentive to develop technologies that minimize the cost of resources needed to produce a good. This is but another way of saying that competition rewards efficiency and maximizes the output that a society is able to produce with a given resource base.

Up to this point the discussion has been of a conceptual nature. Capitalism has been analyzed as an abstract system whose results are deducible from its preconditions and assumptions. These can never be fully realized because the structural preconditions cannot be implemented completely. However, each society does make choices about whether to accept or reject, approach or avoid the capitalist economic model. The choices are manifested in the political system which, at once, reflects and implements societal values.

CAPITALISM IN AMERICA: The Constitutional Framework

As enlightened men of the eighteenth century, our Founding Fathers were committed to democracy and liberty. Spirited debates at the First Constitutional Convention indicate that the delegates found these twin goals to be potentially incompatible. Democracy was defined as ". . . a system of government which directly expressed the will of the majority of the people, usually through such an assemblage of the people as was possible in the small area of the city-state."[6] Thus, democracy was linked to a concept of suffrage whereby the majority of citizens determine the laws that govern all. The concept of liberty, on the other hand, was associated with property. Individual freedom was considered to be grounded in private property rights. As long as an individual could be assured of a right to acquire and hold property, he or she was free to pursue self-interest, enjoy the rewards of industrious activity, and build a stable life.

In the eighteenth century, humans were generally regarded as grasping creatures who pursued personal gain without regard to the welfare of others. Adam Smith accepted this view in *The Wealth of Nations* and argued that the pursuit of individual self-gain in the marketplace would lead to the economic common good because the greed of sellers would be constrained by the greed of their rivals and of buyers. Competing greed would operate as an "invisible hand" to force all traders to produce and sell at the lowest price and thereby utilize resources most efficiently.

However, delegates to the Constitutional Convention understood that the pursuit of self-interest in the political sphere might well nullify its pursuit in economic activities. If all persons were given the vote under a democratic system, the greedy rabble might elect to expropriate the land and goods of the propertied classes. In such a case, there would be no incentive for individuals to acquire resources and allocate them toward profitable uses. Hofstadter summarizes the dilemma as follows:

> Freedom for property would result in liberty for men—perhaps not for all men, but at least for all the worthy men. Because men have different faculties

and abilities, the Fathers believed, they acquire different amounts of property. To protect property is only to protect men in the exercise of their natural faculties. Among the many liberties, therefore, freedom to hold and dispose [sic] property is paramount. Democracy, unchecked rule by the masses, is sure to bring arbitrary redistribution of property, destroying the very essence of liberty.[7]

In 1787 the ownership of property was widely dispersed among the population, but the Fathers were concerned that a democracy might crush liberty if an unpropertied class were to develop at some future time. The Convention consciously addressed the problem by adopting a federal form of government with a Constitution that safeguarded individual liberty. It would stretch the point to say that the Founding Fathers consciously chose to create a political system that assured the existence of a free enterprise economy. Their own economic philosophy was probably closer to mercantilism than to capitalism. Nevertheless, the individual rights that the Founders chose to guarantee either established or protected most of the preconditions necessary for a capitalist economy.[8]

It is clear, from the legislative history of the Constitution, that the authors intended to protect the right of private property. An explicit statement to this effect was added three years after ratification. The Fifth Amendment provided that: "No person shall . . . be deprived of life, liberty or property, without due process of law; nor shall private property be taken for public use, without just compensation."

Article II (Sections 8 and 10) addressed contract enforceability and the establishment of a recognized medium of exchange. The Congress was authorized to coin money, regulate its value, and punish counterfeiters. It was also empowered to establish uniform bankruptcy laws that would apply throughout the nation. On the other hand, states were expressly forbidden to coin their own money or to pass any law that impaired the obligation to fulfill contracts. Charles Beard notes that the establishment of uniform legal tender and the prohibition of state interference in contractual arrangements assured an important measure of stability in commercial transactions:

Contracts are to be safe, and whoever engages in a financial operation, public or private, may know that state legislatures cannot destroy overnight the rules by which the game is played.[9]

The protection of property rights, freedom of contract, and the establishment of legal tender provided the essential cornerstones of a free enterprise system because they assured that goods and resources could be held or exchanged freely in accordance with the best interests of the owner. The Constitution also protected the mobility of people, goods, and resources from governmental barriers. Article IV (Section 2) entitled citizens of each state to all the privileges and immunities of citizens throughout the United States. Article I (Section 9) forbade states to impose any tax on articles exported to other states. Individuals could migrate from state to state secure in the knowledge that they would maintain their full rights as citizens. Resources and goods could be moved in response to economic opportunities anywhere in the nation without being subject to penalty.

The Constitution safeguarded and encouraged the free exchange of information. The First Amendment guarantee of a free press had important economic as well as political and social implications. No governmental limits could be placed on the ability of the news media to satisfy the market's demand for knowledge. Information regarding trade opportunities, employment situations, wages, prices, new products, and technological improvements could be dispersed wherever economic units and consumers found it in their self-interest to send or receive it. Article I (Section 8) provided for post offices to carry this information. It also spoke to commercial and consumer interests by empowering Congress to fix the standard of weights and measures which would provide sellers and purchasers with a uniform language through which they could communicate important terms of trade.

The Constitution failed to address only two of the capitalist preconditions: freedom of prices to fluctuate without government interference and the enforcement of a market structure wherein no industry would be dominated by a single firm or combination of firms. However, the very absence of regulation left prices free to rise or fall in accordance with consumer preferences. Antitrust legislation was not necessary to enforce competition in an economy composed of farmers, artisans, and small manufacturers. Moreover, the nation's enormous land mass, coupled with its rapidly increasing population, produced a marketplace so vast that no industry could be dominated even by an association of firms.[10]

It is important to notice the absence of any constitutional provision requiring entrepreneurs to obtain a federal charter in order to conduct business. Legislation would not be used to create national monopolies or limit the entry of economic units into industries and occupations. States remained free to require charters and certificates, but market forces produced competition among the states that sharply limited the use of this power to "industries clothed with the public trust" and natural monopolies, e.g., utilities.

It is true that certain constitutional provisions provided Congress with explicit powers that had the potential to curtail free enterprise. The right to levy and collect taxes might be exercised in a manner that would expropriate profits and destroy incentives. The Commerce Clause might become a basis for legislation forbidding the sale or the production of goods that were deemed "not in the public interest." The appropriate limits for these and other Congressional powers relating to economic activity would be the subject of argument throughout the nation's lifetime.

In summary, the Constitution created a political structure that was entirely compatible with a capitalist economic system. It established or protected almost all of the preconditions for free enterprise and, in so doing, safeguarded the ability of individuals to pursue their economic self-interest within a market economy.

CAPITALISM AND MORAL VALUES

The first section of this essay argued that a society can be stable only if its economic system satisfies material and moral criteria. Few would deny that American free enterprise has met the material test of creating large aggregate wealth for the nation. But has it met the moral test? The question is difficult to answer. On the one hand, it is doubtful whether any economic system is capable of satisfying every ethical criterion. On the other hand, the very fact that—with the exception of the Civil War—our nation has maintained a relatively stable society over the past 200 years is *prima facie* evidence that our economic organization has met at least the minimum consensus standard of morality.

Perhaps it is more fruitful to address the moral issue with a different set of questions. Does capitalism, considered as an abstract economic system, necessarily impose moral values of any kind? To what extent has American public policy shown itself able to address moral problems, either by enforcing competitive preconditions or by making selective modifications in our economic system?

Private property, contract enforceability, and the pursuit of self-interest are fundamental to capitalism. Philosophers and economists have, for centuries, debated the ethical merits of these preconditions. Considerations of space do not permit even a summary of the arguments here. It is desirable, however, to state the propositions regarding these concepts that must be acceptable if capitalism is to be capable of ethical justification.

Individuals exercise a proprietary claim to resources by "operating" on them in a manner that produces goods and services. Unless such a property claim is permissible, resources will not be developed to their fullest potential for serving the common good.

Individuals have a moral obligation to meet the terms of agreements that they have made with others. Therefore, societies of individuals are morally justified in enforcing such terms.

Individuals are best able to determine self-interest and are morally justified in pursuing it. The "common interest" is in some way related to individual interest.

With these propositions stated, but not argued, the discussion turns to a consideration of moral issues in the dynamics of capitalism.

The discipline of economics has traditionally emphasized the efficiency aspects of capitalism. Microeconomic analysis demonstrates that the resource costs necessary to produce any good are minimized when the preconditions are strictly fulfilled. In analyzing competitive dynamics, economists generally make no "value judgments" regarding the nature of the goods or of the production process. Their singular emphasis on cost minimization and failure to address value questions (before or after a "value-free" analysis) have, unfortunately, resulted in two widely shared misunderstandings about the relationship between moral values and the dynamics of capitalism. Some have concluded that the market system is itself "value-free" and that moral values cannot be expressed in a capitalist economy. Others have concluded that competition to minimize costs inevitably drives out ethical practices and results in worker or consumer exploitation. Both views are incorrect.

The very nature of capitalism allows the widest possible expression of ethical norms. It permits moral as well as material pluralism because no universal ethical code is imposed by *fiat* or even by informal consensus. Rather, the market reflects the norms of the individuals and firms that operate within it. The preservation of competition enables consumers, investors, and employees to apply their personal ethical systems to market activities by dealing only with the economic units whose behavior they find appropriate. When many suppliers are available, consumers need not buy from firms that discriminate, pollute the environment, exploit employees, or engage in other immoral activities. Competition assures diversity, and diversity assures the ability to select among firms on the basis of ethical principles as well as product price and quality.

If the preservation of a given ethical norm is important to the satisfaction of a set of consumers, then the market will bring forth economic units to produce goods subject to the constraints of that norm. Of course, the consumers may have to pay some incremental price if their moral standards conflict with minimum cost technology. But that incremental price will equal only the cost of the additional resources that the producer expends in using some alternative technology that is preferred on ethical grounds.

With respect to the second misunderstanding, competition can enforce as well as erode ethical practices. To the extent that the capitalist preconditions are fulfilled, there is widespread knowledge of prices, wages, product quality, and working conditions. Consumers will be drawn to the goods that are most satisfying, and firms will be driven out of business if they fail to meet the standards of price and quality that prevail in the market for their products. Similarly, workers will avoid firms that attempt to undercut competitive wage rates or working conditions and accept employment that meets the market standard. In so doing, they set in motion forces to destroy exploitive conditions. Thus, a firm's ability to minimize prices and costs is constrained by the actions of market rivals. Competition sets a floor as well as a ceiling upon the quality of goods, conditions of the work place, and wages.

These "protective" processes can operate effectively only if consumers are concerned about the ethical implications of their choices and are informed about market alternatives. Otherwise, competition will operate without moral constraint as it drives costs to their minimum levels. Competition's moral results are ultimately determined by the value system of consumers. Whatever the nature of such values, individuals must have information about products and company behavior if they wish to deal with firms that adhere to a particular ethical standard and avoid those that do not. A vigorous, free press is essential for providing the information necessary to exercise these choices.

CAPITALISM AND AMERICAN PUBLIC POLICY

The preconditions of capitalism can never be fully attained in a world of imperfect humans and imperfect institutions. To the extent that the implementation of capitalism departs from its abstract ideal, the market's ability to satisfy the material and moral desires of consumers will not be realized. Some imperfections may be created if economic units exercise their freedom of contract to form cartels. Others may result, at least temporarily, from inadequacies in the nation's information and transportation systems. A few are inherent in the nature of certain industries. All such imperfections reduce the economic system's ability to allocate resources efficiently.

Even if the preconditions of capitalism could be fully achieved, the resulting distribution of goods might not meet society's ethical standards. Capitalism rewards economic units, including workers, in accordance with their productivity. It provides no income opportunities for those who lack personal or material resources that command a price in the marketplace. Few persons would find moral satisfaction in an income distribution system that only gives "to each according to his ability." Most feel some obligation to provide for those who, through no fault of their own, are unable to fend for themselves in the marketplace. Others would at least agree that social stability is threatened unless this obligation is met.

Charitable contributions to private income transfer agencies are one approach to the problem. However, historical experience suggests that voluntary giving is likely to be an inadequate and uncertain system for providing a minimum living standard, consistent with society's overall wealth, to those in need. In the absence of a strong cultural norm or other informal enforcement mechanism, individuals may agree in principle that a private transfer system is desirable—or necessary—and yet find it in their own self-interest to let others provide all of the support. In so doing, they would gain the benefits of increased social stability but bear none of the cost. The dual considerations of moral obligation and social benefit argue for augmenting private charity through a transfer system supported by mandatory contributions.

American society has addressed these issues through its political system by selective public policy to affect the allocation of resources and distribution of income. Except for wartime legislation and the brief flirtation with wage and price controls during the early 1970's, legislative interventions in the markets for material resources and goods have generally attempted to enforce competition or "perfect" the capitalist preconditions that are grounded in the Constitution. However, public policy has followed a "mixed strategy" in the labor market by specifically departing from some capitalist preconditions and enforcing others. The nation has also chosen to supplement capitalism's distribution systems with mandatory income transfers financed by taxes.

Enforcement of Competition. During America's first 100 years, the economy grew from a series of disaggregated markets served by local producers to a national market populated by industrial giants. The expansion in the absolute size of firms often brought benefits to the consumer because economies of scale combined with vigorous competition to yield low costs and prices. In several basic sectors, however, the growth was also marked by increased concentration of production within a dominant firm or a cartel. This was particularly true of the railroads, sugar, petroleum, dressed meat, and tobacco industries. Monopoly power was exercised to divide markets, restrict output, foreclose entry of new firms, and drive prices upward.

There was a widespread reaction against the huge profits earned by the "trusts" and the inordinate power exercised by a few industrialists. In the election of 1888, both parties courted the vote by promising action against the trusts. After his victory at the polls, President Harrison called for antimonopoly legislation and Congress responded in 1890 with the Sherman Act. The statute is a model of simplicity:

> *Section 1.* Every contract, combination in the form of trust or otherwise, or conspiracy, in restraint of trade or commerce among the several States, or with foreign nations, is hereby declared to be illegal.
>
> *Section 2.* Every person who shall monopolize, or attempt to monopolize, or combine or conspire with any other person or persons, to monopolize any part of the trade or commerce among the several States, or with foreign nations shall be deemed guilty of a misdemeanor. . . .

The goal was to ensure that industries would be populated by several competitive firms. It limited freedom of contract by outlawing agreements to monopolize an industry. In so doing, the Sherman Act protected consumer welfare and preserved the freedom of economic units to enter profitable ventures.

In 1902, Theodore Roosevelt used the Act to prosecute the Northern Securities Company on a monopoly charge. This holding company had been organized by J. P. Morgan and consisted of a gigantic railroad network embracing the Northern Pacific, the Great Northern, and the Chicago, Burlington, and Quincy Railroads. In a 5–4 decision, the Supreme Court found in favor of the government. Although this was not the first successful exercise of the Sherman Act, the magnitude and infamy of Northern Securities made it the most dramatic of the early cases. Later convictions followed against the tobacco, meat packing, and oil cartels. Enforcement was particularly vigorous under Woodrow Wilson. During his term of office the reach of the Sherman Act was expanded with the passage of the Clayton Act.

The evolution of antitrust legislation has continued well into the present century. By enforcing a diffusion of economic power throughout each industry, it protects the welfare of consumers and provides entrepreneurs with an opportunity to enter profitable industries. This body of law, as interpreted by major Supreme Court decisions, is a reaffirmation of the nation's basic commitment to a competitive system.

The Civil Rights Act of 1964 provides a second major example of legislation that addresses moral and economic problems by enforcing competition. Title VII forbids employers, employment agencies, and labor organizations to discriminate against any individual with respect to compensation or conditions of employment because of race, color, religion, sex, or national origin.[11] The underlying policy is entirely consistent with the capitalist ideal. As interpreted by the Supreme Court in the landmark case of *Griggs* v. *Duke Power*, the Act requires employers to hire, promote, and compensate workers strictly on the basis of job performance and productivity. It prohibits the consideration of specific personal characteristics that are unrelated to the job and requires economic units to follow the efficiency criterion in employment decisions.

The Civil Rights Act is inconsistent with the capitalist ideal to the extent that it limits the free choice of employers and—in the case of personal services—consumers. However, historical evidence had shown that an unlimited exercise of contractual rights by some individuals had infringed upon the freedom of employment contract for others, with the result that the latter group could not achieve the competitive promise of being rewarded in accordance with economic contribution. As in the case of antitrust laws, the restriction of some contractual rights was necessary to achieve the competitive goal of free access to economic opportunity.

Enhancement of Competitive Preconditions. Effective competition in markets for goods and labor requires that consumers, economic units, and workers have reasonable knowledge about prices, product quality, wages, and working conditions. The market itself is able to provide much of this knowledge through the news media, the stock exchange, voluntary consumer unions or cooperatives, and private employment agencies. At various times in the nation's history, however, some firms have engaged in deceptive or even fraudulent practices to exploit consumers or employees. Although many of these activities were actionable at common law, the public favored the creation of legislation to guide the courts.

The history of the Pure Food and Drug Act exemplifies this branch of public policy. In 1906, Upton Sinclair wrote *The Jungle*, a novel which graphically depicted the filthy conditions in meat packing plants, the adulteration and sale of putrified food, and the industry's disregard for the health of consumers. When a government investigation substantiated the conditions described by Sinclair, Congress was swift to pass the Pure Food and Drug Act. Samuel Eliot Morrison notes that producers fought strongly against this "socialist" bill. He quotes a candy maker who had added shredded bone to coconut bars as saying, "It don't hurt the kids, they like it!"[12] The Act was the first in a series of legislation that requires business firms to label products accurately, provide health warnings where appropriate, and maintain minimum standards of truth in advertising.

There would be no need for such protective legislation if competitive preconditions could be fully implemented. Consumers would have full knowledge of all product characteristics and could make intelligent choices among goods with respect to quality and safety. The passage of the Pure Food and Drug Act and its successors reflects the nation's experience that the marketplace does not necessarily generate the required levels of knowledge to permit consumers to make an informed choice within some product categories. Laws that require the diffusion of such information are consistent with the capitalist system. In some instances, legislation has gone beyond the point of assuring that consumers are informed either by forbidding the sale of certain goods or mandating the inclusion of specific safety features. (The short-lived "interlock" seat belt system and the Laetrile controversies are examples.) Such interventions are subject to serious criticism because they prevent consumers from exercising an informed choice.

Public policy has also enhanced the competitive preconditions of knowledge and mobility in the labor market. The Wagner-Peyser Act of 1933 created a federal-state employment service to function as a clearinghouse for job information. Employers list detailed descriptions of open positions that can be matched against the skills of applicants. The system operates voluntarily and without charge to either employers or applicants. It is financed by a tax that is ultimately passed on to consumers. The underlying policy of this free service is to maximize the labor market choices of firms and workers.

It might be argued that the creation of a publicly supported labor exchange is inconsistent with the nation's commitment to free enterprise because the value of labor market information is so high that employers and job seekers would be willing to purchase it from private agencies operated for profit. The argument is generally answered in terms of the positive "neighborhood effects" provided by the service. An efficient labor market that minimizes the length of unemployment and the number of unfilled positions yields benefits that accrue to society at large as well as to the particular individuals and firms that utilize the employment service. Those who receive the neighborhood benefits should bear a portion of the cost. All beneficiaries cannot be individually identified, and their widespread dispersion throughout the economy argues for supporting the system through tax dollars.

Economic activity produces neighborhood costs as well as benefits. This has given rise to public policy that enhances competition by requiring firms to pay for all of the resources used in production and marketing. The classical economists recognized that even a perfect competitive system would not require firms to pay for the cost of some resources, such as water and air, that are necessarily held in common. In the absence of economic or legal constraints covering all producers, the profit incentive would induce economic units to gain a competitive edge by adopting techniques that make extensive use of such "free" resources. In so doing, they would shift part of the resource costs to the public through neighborhood costs such as pollution, excessive noise levels, and erosion. This damage to the rights of others can be reduced by legislation that "taxes" firms for the use of public goods. In recent years, the nation has developed such laws as the Clean Air Act to monitor effluents and charge firms for the damage done to the environment. The costs are ultimately passed on to the consumer in the form of higher prices. This is entirely consistent with the capitalist system which—if perfected—would impound all resource costs in the price of the final good.

Departures from competition. There are relatively few instances in which the nation has chosen to depart from competition in the market for goods. A major exception has been the substitution of regulation for competition in industries characterized by declining average cost over a substantial range of output. These "natural monopolies" include utilities, radio-television, and common carrier transportation. The argument for replacing competition by regulated monopoly is technical rather than ethical, and will not be treated here.

Public policy in the labor market offers the most significant examples of legislated departures from competition. Most workers have a legally protected right to form cartels. Unions are associations of workers that seek to increase wages and improve working conditions by concerted action that includes collective negotiation, joint withholding of labor (strikes), joint refusals to purchase certain goods (boycotts), and restrictions on entry into certain occupations. The antitrust laws forbid firms to conduct such concerted activities in the product market. Early Supreme Court decisions found that the Sherman Act also applied to union activities; boycotts and secondary strikes were held to be restraints of trade under the Act. Disruptions of interstate sales by means of strikes against primary employers also appeared to be reachable under the Act.[13]

The American Federation of Labor undertook a major campaign to exempt unions from the Sherman Act. The effort achieved success with the inclusion of Section 6 in the Clayton Act of 1914:

> . . . The labor of a human being is not a commodity or article of commerce. Nothing contained in the antitrust laws shall be construed to forbid the existence and operation of labor . . . organizations . . . ; nor shall such organizations, or the members thereof, be held or construed to be illegal combinations or conspiracies in restraint of trade, under the antitrust laws.

The full judicial impact of this remarkable provision was not realized until the Apex Hosiery decision of 1940.[14] By that time, Congress had already passed the Wagner Act (1935), which went beyond the antitrust exemption to provide workers with a legally protected right to engage in concerted activity. Section 7 stated:

> Employees shall have the right to self organization, to form, join, or assist labor organizations, to bargain collectively through representatives of their own choosing, and to engage in concerted activities, for the purpose of collective bargaining or other mutual aid and protection.

Other provisions of the Act forbade employers to interfere with the exercise of employees' rights under Section 7 and outlawed specific antiunion practices. There have been two major revisions of the Wagner Act, but the nation has retained the basic policy of protecting workers' rights of self-organization and exempting from antitrust laws union activities that do not involve collusion with employers.

Congress justified this departure from competition by the theory of countervailing power and the need for industrial peace. In Section 1 of the Wagner Act Congress found ". . . [an] inequality of bargaining power between employees who do not possess full freedom of association or actual liberty of contract, and employers who are organized in the corporate or other forms of ownership association. . . ." It also found that commerce was burdened by strikes which resulted from the denial of employee organization rights and the refusal of some employers to accept collective bargaining. Thus, the bases for our labor market policy are partly structural and partly empirical. The countervailing power argument is grounded in the belief that the preconditions of competition cannot be fulfilled in the labor market to the extent necessary to protect employee welfare. Imperfections in labor market information, mobility, and market structure are presumed to provide firms with monopsony power to drive wages and working conditions below competitive levels. The nation has elected to adopt a "second best" solution of offsetting power with power. This policy has not entirely freed organized workers from market constraints. The Supreme Court has consistently held that employers may replace workers who are striking purely for economic reasons.

The legislative policy of protecting worker associations was also based upon a concern for industrial peace. The history of labor unions in America provides ample evidence that restrictions on concerted activities by workers are likely to trigger work stoppages and even violence. The perception that working conditions are unfair generates a cohesion among employees that logically expresses itself in coordinated action. The adoption of a public policy to prevent such activity is certain to encroach upon perceived freedoms of association and speech. It may well be the case that acceptance of the second best solution is necessary for preserving the stability of our social system.

The protection of worker organizational rights has been accompanied by comparative peace in labor relations. Since 1947, the fraction of total worker days lost to strikes has averaged .2 of 1 percent. Although most workers have a protected right to organize, the vast majority have not exercised the option. Approximately 21 percent of the labor force is covered by a collective bargaining agreement. Earnings and working conditions for most of the others are set by market forces.

Other departures from competition in the labor market include the passage of protective legislation relating to occupational safety and health, child labor, maximum hours, and minimum wages. All of these limit the freedom of contract between employers and employees by forbidding certain terms of employment. The minimum wage law involves a particularly interesting value judgment in public policy. By imposing a floor on wages, the Fair Labor Standards Act specifically departs from the competitive precondition of freely fluctuating prices and wages. In setting the minimum wage above the market levels, the Act runs a very real risk of causing employers to fire those workers whose productivity does not meet or exceed the mandatory rate. Unfortunately, this unemployment effect is most likely to fall on individuals who are disadvantaged in terms of education, skill, and age. Thus the minimum wage law thwarts the very goal it seeks to achieve.

Income transfer systems. In a capitalist system, the income of individuals is determined by the market value of their personal and material resources. Such a distribution system would raise few ethical issues if all persons possessed the resources necessary to earn at least a minimum living standard. In such an ideal world, individuals would starve only if they freely chose not to employ their talents. However, many persons do not possess marketable resources and, through no fault of their own, are unable to earn a market income. There is a strong moral case to be made for providing them at least minimal support.

American public policy has addressed the issue by permitting charitable donations to be deducted from taxable income. This approach is clearly consistent with a capitalist system. The nation has also created a series of income transfer programs financed by taxes. Persons who are able to compete in the marketplace have a portion of their income taxed and distributed to those who are temporarily or permanently unable to make a living. Usually, recipients are screened by some form of "means" or "effort" test to verify that they lack personal resources or have failed in their best attempt to become employed. Federal programs transfer income to needy persons who are disabled, aged, or members of families with dependent children. In addition, a system of unemployment compensation has been developed to provide temporary payments to eligible individuals who have been previously employed, have involuntarily lost their job, and are actively seeking work.

There are a host of ethical problems raised by the manner in which some of these programs have been implemented.[15] These complex issues will not be addressed here. However, it is important to state that the creation of an income transfer system is not inconsistent with a capitalist economy unless the tax burden becomes so heavy that it destroys the profit incentive for economic units. It is obvious that society cannot transfer more income than it creates and that potentially productive individuals will refrain from entering the market if their nonwage income alternatives are sufficiently attractive or if their income, net of taxes, does not provide adequate compensation for their effort.

CONCLUSION

Individual liberty is the foundation of free enterprise. The pursuit of self-interest within the framework of capitalist preconditions produces an efficient allocation of resources toward the products and services desired by consumers. The system operates with a minimum of coercion. Individual consumers freely determine their own needs; economic units freely choose the needs they will serve and the manner in which they will serve them. The process permits the widest possible diversity and creativity. The only limit on the quantity, quality, and range of goods produced is the size of the profit opportunity and the extent of the market. There is no central authority that mandates the allocation of resources to satisfy "the common good"—however determined.

Capitalism also provides a mechanism for assuring that individuals will be responsible for their own economic actions. Its preconditions impose the discipline of the marketplace that constrains behavior even as it gives the broadest scope to the pursuit of self-interest. An economic unit earns a profit or suffers a loss to the extent that it serves or fails to serve consumer desires in an efficient manner. Each consumer pays the full cost of all privately owned resources used in the production of the goods he purchases; the cost of those resources is set by the price that other consumers are willing to pay for their use.

By preserving liberty and assigning responsibility, capitalism permits individuals to shape their economic behavior in accordance with personal moral standards. In a free society, some diversity of norms is assured because moral values are shaped by a complex interplay of educational, religious, and sociological factors. The capitalist principle of consumer sovereignty assures that plural values may be expressed in economic activity. Free enterprise is entirely compatible with the highest systems of ethics provided that individuals express those systems in the marketplace.

There is no question that the implementation of free enterprise must fall short of the abstract ideal because its preconditions can never be fully satisfied. Selective government intervention is therefore necessary. It is the function of economic public policy to guide the system toward a more complete achievement of its goals. In most cases, America has addressed economic issues of moral significance by protecting the rights of individuals to seek self-interest within a competitive system. This has been expressed in legislation that prevents restraints of trade, protects equal opportunity in employment, improves market information, and requires producers and consumers to pay the neighborhood costs for the use of resources that are held in common. The major departure from capitalist preconditions has occurred in the labor market, where legislation permits workers to form cartels and sets limits upon the freedom to contract for certain working conditions. This "second best" solution was adopted because imperfections in the labor market seemed beyond legislative resolution. It also reflected a concern for industrial peace. Finally, the nation has developed a system of public transfer payments to provide income for those who cannot compete.

Our Founding Fathers recognized that public policy set by majority vote might destroy liberty. Government regulation inevitably encroaches on individual freedom. It is inflexible and relatively unconstrained by efficiency considerations. Thus, economic legislation must be used selectively and sparingly if the nation is to retain the benefits of capitalism. Solutions to moral problems should be sought in the private sector before public initiatives are undertaken. When private measures cannot be found, preference should be given to legislation that addresses such problems by enhancing or enforcing capitalist preconditions. Administered solutions that limit or destroy market forces should be adopted only as a last resort.

Capitalism, as implemented in the United States, has proven to be a robust economic system. It has produced—and survived—a remarkable industrial expansion within a nation of rapidly growing population. It has weathered dramatic shifts in social norms and shown itself capable of responding to moral issues as they have been identified. This resiliency is grounded in capitalism's ability to permit the direct expression of individual needs and values without requiring the achievement of consensus by authority or complex social process. Ultimately, then, the direction of this system will be determined by individual decisions.

NOTES

1. The terms "free enterprise" and "capitalism" are used synonymously throughout this essay.
2. The term "goods" includes services as well as material products; "resources" includes human as well as material factors that may be used to produce goods. An "economic unit" is any person or entity that buys or sells resources.
3. "Cartels" include economic combinations of workers, i.e., unions as well as firms.
4. Of course, price differentials will be arbitraged only if they exceed the costs of transportation between the two locations.
5. For a more complete discussion of the effects produced by an absence of competitive conditions see George J. Stigler, *The Theory of Price* (New York: The Macmillan Company, 1952), Chapters 12–14.
6. Richard Hofstadter, *The American Political Tradition* (New York: Vintage Books, 1948), p. 12. This section draws heavily upon Chapter 1 of Hofstadter's work.
7. *Ibid.*, 11 and 12. Footnote omitted.
8. The discussion of the Constitution and capitalist preconditions is consistent with the treatment in Charles A. Beard and Mary R. Beard, *A Basic History of the United States* (New York: Doubleday, Doran, & Co., 1944), Chapter IX.
9. Charles A. Beard, *An Economic Interpretation of the Constitution of the United States* (New York: The Macmillan Company, 1913), p. 179.

10. The United States included 880,000 square miles of land in 1790. Subsequent purchases, treaties, and cessions increased these holdings to 3.6 million square miles. Steady immigration caused the population to grow from 4 million in 1790 to 17 million in 1840.
11. Section 703(e) allows religion, sex, or national origin to be considered in those instances where they are a ". . . *bona fide* occupational qualification reasonably necessary to the normal operation of that particular business or enterprise. . . ." Court decisions have interpreted this section very narrowly.
12. Samuel Eliot Morison, *The Oxford History of the American People* (New York: Oxford University Press, 1965), p. 819.
13. *Coronado Coal Co.* v. *UMW* 268 U.S. 595 (1925).
14. *Apex Hosiery Co.* v. *Leader* 310 U.S. 469 (1940).
15. See, for example, Frances Fox Piven and Richard A. Cloward, *Regulating the Poor* (New York: Vintage Books, 1971).

II

IS FREE ENTERPRISE COMPATIBLE WITH SOCIAL JUSTICE?

3

CORPORATE COLLECTIVISM:

A System of Social Injustice

MICHAEL HARRINGTON *

I do not believe the United States is a free enterprise society. I think the people who want us to use this term wish to rationalize and defend rather than describe the society. There may have been a free enterprise society for five minutes in Great Britain in the nineteenth century, although personally I doubt it. It certainly has never existed in the United States. Among other things, we fought a war, the Civil War, in which one of the issues was free trade versus protectionism. Protectionism won. The infrastructure of our society, the railroad, was built by federal donations. The United States today does not have a free enterprise society in any kind of Adam Smithian sense of the term. Rather, we have gigantic oligopolies administering prices, shaping tastes, working together with an all-pervasive government which follows corporate priorities. Therefore, I suggest that rather than using the term "free enterprise" to describe our system we get closer to reality and call it "corporate collectivism."

I maintain that corporate collectivism, in its historical thrust and tendency, is not compatible with social justice. Moreover, I am absolutely certain that capitalism—corporate collectivism in its latest phase—is coming to an end. As a socialist, I am not necessarily filled with joy at this prospect because there is more than one end possible to this society. The question before us is not *whether* there is going to be a collectivist society in the future, because that is already decided. The real question is, what kind of collectivist society will there be? When we play around with terms like "free enterprise," we ask the wrong questions and cannot possibly come up with the right answers.

* This paper is based on the transcript of a talk given at a symposium held at the University of Kansas in November, 1976.

I will approach these issues from three angles. First of all, I shall look at capitalist society in its classic and heroic phase, and give a very procapitalist analysis, because that has always been part of the socialist case. I want to do so not simply to praise capitalism, but to understand its origins as they impinge on the second point, contemporary capitalism. That will allow me to proceed to the third point, the alternatives to contemporary capitalism. In short, I shall touch on the past, the present, and the future of capitalist society in the United States or, for that matter, in the world.

I. CAPITALISM: An Historical Perspective

First of all, the capitalist past. Capitalist society is the greatest human society that has ever existed. It started in the West, of course, and brought to the world a number of enormous benefits. I state this without any hint or trace of irony. The capitalist revolution saw the spread of science and rationality. It gave democracy to the world. It gave the world a concept of individual worth that did not exist on anything like a mass basis in any precapitalist society.

Nonetheless, my enthusiasm should not be taken as too procapitalist. One of my criticisms of the *Communist Manifesto* is that Marx is excessively procapitalist, that he gives the capitalism of his age credit for prodigious things which it was not to accomplish for another fifty years. In talking about the accomplishments of capitalism, I also want to recognize its Hobbesian reality, which is so marvelously described in the *Leviathan* and its Darwinian struggle of each against all. The capitalist revolution was not pretty; yet on balance, it was one of the most magnificent advances that humankind has ever made.

Now, I want to understand that advance. What was the principle of precapitalist society? What was the big gulf between capitalist society and previous societies? In all that went before capitalism, the political, the economic, and the social were one. The individual's economic position in society was not independent of his political and social position. If you were a duke, a slave owner, or a member of a mandarin bureaucracy, there was a seamless whole which defined you politically, economically, and socially. In the Middle Ages, a young serf did not decide to try to become a duke. There was only one institution in the Middle Ages which made that possible—the Catholic Church. But, by and large, it was a society where one's economic, political, and social position was fixed at birth and was part of this whole.

As a result, in these precapitalist societies (what Marx called Oriental despotism, and slave society, as well as feudal society), the ruling class had a very specific way of extracting an economic surplus from the direct producers. Feudal society achieved the surplus through force sanctioned by religion. That is to say, the serf labored free for the lord or gave him rent in kind, or later gave him money, because the lord was his military protector, sanctioned by God, in an order which was accepted from the very start.

Capitalism did a number of absolutely magnificent things in terms of those precapitalist societies. First it separated, or as a system had a tendency to separate, the political, economic, and social orders. Properly speaking, social classes are an invention of capitalism. Before capitalism, there were "estates," caste systems, hereditary systems. Only with the advent of capitalism and its separation of the political, economic, and social systems was there a society that offered individual choice, the possibility of mobility, and an opportunity to rise above the status of one's birth.

Second, capitalism normally did not extract a surplus by naked force. That was a great advance. Capitalism is the system which uses economic means to get the surplus out of the direct producer. That is to say, it is based on a contract. It is absolutely true that the contract is utterly unfair, but it is nevertheless a contract, and that means a rise, so to speak, in civil liberties—a rise in the ethical and personal level of the society. The contract is unfair because, in the capitalist mythology, a free worker with nothing to sell but his labor freely contracts with someone who has enormous wealth and who is able to gain a surplus through that contract. It is an unfair contract, but it is a system that works on the basis of contract and law rather than force. It is a system which prefers to organize itself democratically. There are times when this system will leave democracy—the Germany of Adolf Hitler is one of the most monstrous and horrible examples—in a period of crisis. But normally, capitalist society prefers to function democratically with certain civil liberties. That is the second enormous contribution to society.

Third, capitalism discovered the social power of people working together. That, I submit, was its economic invention. Capitalism did not begin with a technological revolution. Indeed, capitalist technology—the technology of the sixteenth and seventeenth centuries during the origins of capitalism—was, on the whole, conservative and was borrowed from other cultures. We borrowed the compass, gunpowder, and the printing press. We borrowed something of enormous importance, Arabic numerals, from the Arabs, who had taken them from the Chinese. A modern society cannot be built on the basis of Roman numerals;

it needs the zero concept and positional notation. We were technologically, perhaps, somewhat behind Chinese and Byzantine societies of the fifteenth century. What capitalism discovered was not technology; it discovered that allowing people to work together enormously increased their productivity. That was social invention. The discovery of the social nature of work, of people working together rather than in isolation, was the decisive moment in the rise of capitalism and, again, an enormous contribution to humankind.

This brings me to the final point of this brief history: the capitalist system bases itself on a contradiction. It is a system which is increasingly and progressively social capitalism. Capitalism is a socializing system. It first socializes work, with entrepreneurs getting artisans together under a single roof, achieving certain economies of scale. Larger entrepreneurs socialize smaller entrepreneurs. There is a tendency toward concentration, toward monopoly, toward oligopoly, cartels, trusts—larger and larger units. Capitalism develops a revolutionary technology, with social consequences permeating the entire society. Henry Ford obviously is one of the great radicals, one of the great revolutionaries of modern life.

Capitalism is from the very first moment a world system. It reaches all over the world, destroys ancient cultures, and upsets the international balance. It tries to bring the entire world, every last person, into the same integrated unit. Capitalism is profoundly social and profoundly revolutionary. But at the same time that capitalism becomes more social, at the same time that the technological decisions today of a U.S. Steel or an IBM have more social consequences than most of the decisions of state legislatures, the system remains private in its decision making and its allocation of resources and benefits. And there, I suggest, is the genius of capitalism and the contradiction that will destroy it. Here I am using a paradox defined by Joseph Schumpeter, one of the great conservatives of the modern age, but I think it very much in the spirit of my good friend, Karl Marx, viz., that capitalism is destroyed by its success, not its failure. Capitalism so successfully socializes that the socialized world which it creates becomes incompatible with the institutional decision-making process in private hands.

Finally, I am perfectly aware that capitalism today is no longer run by individual entrepreneurs making personal decisions. I am perfectly aware that there are huge corporate bureaucracies, technostructures if you want. Of course, corporate bureaucracies tend to view profit maximizing on a much longer time span than a nineteenth century robber baron. But what I am saying is, for example, that the American automobile industry still considers the problem of pollution or the problem of automobile safety from the point of view of making as much money in a sophisticated way as it possibly can. This is in contradiction with the extremely social system which capitalism creates.

So my first point is this. Capitalism is a magnificent, revolutionary, and liberating system which, by its own success, becomes conservative and reactionary. It will bring about its own demise.

II. CONTEMPORARY CAPITALISM

What about contemporary capitalism? I now move from general theorizing about the nature of capitalism to some current facts. I call our present system corporate collectivism. Its origins are to be found in three historical periods and the events associated with them. The first period was from about 1880 to 1900, when European and American capitalism went through a crisis which it solved by monopolies, trusts, and by the intervention of the state, the beginning of the state's entrance into the economy.[1] The first move toward corporate collectivism is the end of the entrepreneurial capitalism at the end of the last century.

The second period was World War I, a tremendous watershed. Here I use the rather classical, conservative, contemporary economist Hicks as my source.[2] Hicks points out that during World War I, the government had to mobilize the war effort. This brought about the discovery of military socialism, the capacity of the state to administer the economy. There had been government intervention into the economy before World War I, particularly social welfare intervention under Bismarck in Germany and Lloyd George in Britain. But in World War I, government actually operated industries. For example, the railroads were run by the United States government. That was, Hicks says, the beginning of an administrative revolution in which the state learned its capacity for economic intervention.

The third period was during the Great Depression of the 1930's, when Keynesianism triumphed in its most conservative variant. In the 1930's, some nations began to understand that the government had to be responsible for the economic management of the macro-economy. A new role for government was defined in the 1930's, by two countries in particular. One was Nazi Germany, which solved the crisis of the Great Depression by a kind of military Keynesianism, by a fascist planned economy that eliminated unemployment but cost the world the horrors of genocide and World War II. The other practical demonstration of the power of the government to manage an economy was, of course, made by the Swedes, who were the first democratic people to think their way through the problems of capitalism and to solve the Depression of the 1930's. The Swedish people repaid the Swedish Socialist Party for that accomplishment by keeping it in office for forty-four years.

This Keynesian revolution in its democratic variant, particularly in its American variant, was conservative. Many in the United States, including conservatives, think that Keynes was some kind of radical. I suggest they read his life. Keynes was not only a theorist, he was a man who made money for himself and various charitable causes for which he worked as an investment advisor. Keynes was an upper-class aristocratic snob who regarded the workers as a bunch of boobs and who said that he was a Keynesian precisely because he was not a socialist.

Now there are Keynesians who, in my opinion, are much better than Keynes. But American conservatives will have to learn one of these days that the ideology of conservatism in the late twentieth century is Keynesianism or at least conservative Keynesianism. The point of Keynesianism was that the corporate infrastructure was fundamentally sound and that all the government did, merely by fiscal and monetary policy, was to match the macro-aggregates of supply and demand through public and private investment at a full employment level. But the assumption behind Keynes's policy, and particularly behind the Roosevelt New Deal version of Keynesianism, was that the basic investment decisions are made by corporations and that the government simply creates an environment in which they will be able to carry out that function.

So Keynesianism, I suggest, has this profoundly conservative aspect. It has another aspect—an aspect which makes any government, including a liberal government, a hostage to the corporations. I take a very wise sentence from a contemporary German sociologist, Klaus Hoffa: "The capitalist state is not itself a capitalist." That is to say, the capitalist state is dependent for the success of its policies on the private sector. In this society of supposed equals, some people are more equal than others. When a John Kennedy is elected in 1960 or a Jimmy Carter in 1976, they must assure themselves of the cooperation of the corporate community because that community makes investment decisions, and private investment decisions are the key to full employment.

I therefore suggest that the concept of "trickle down," the concept of the government rewarding the corporate rich more than anyone else, is not a policy of American society. It is *the* policy of American society. The policy is imposed upon this society by the fact that corporations are in charge of the investment process. Therefore, if one considers the Kennedy/Johnson tax cuts, as has Leon Keyserling, who was chairman of the Council of Economic Advisors under Truman, one finds that the rich received much more than anyone else. Indeed, I would suggest to you that the welfare state does much more for the corporate rich than for anybody else.

For now, I just want to focus on the fact that macroeconomic policy has to be pro-corporate. If we elected a government in the United States composed solely of left-wing trade union leaders, militant civil rights advocates, feminists and reformers, but kept corporations in charge of investment, that government would be nicer to corporations than to workers or minorities or women or to the people in general—not because it sold out but because that is the reality of government in our society. What I am saying is: the political freedom of society is profoundly restricted by the economic and social institutions. So it is that today in Great Britain a Socialist Chancellor of the Exchequer, a former member of the Young Communist League, is presiding over increasing unemployment and holding wages down as a socialist policy. Under the circumstances in Britain, corporate people are more important than working people; people in board rooms are more important than unemployed people. That is built in, I suggest, to this kind of society.

This leads me to a proposition about the kinds of decisions that are made in our society. I am not suggesting that there is a capitalist conspiracy, that there are a bunch of bloated plutocrats in top hats who sit around the table in the morning and say, "Let us now go and see how we can do harm to the weak and the poor in this society." What I am saying is that there is a structure of power that dictates pro-corporate outcomes to democratically elected representatives.

The transportation industry is an example. Think of what we have done about transportation in this society in recent years. We started a federal highway program under Dwight Eisenhower in the late 1950's. I believe that it is probably, with the possible exception of Medicare, the most important social program of the United States since the New Deal. This program helped the middle class get out of the cities, isolated the cities, helped destroy the railroads, helped destroy mass transportation within the cities, isolated the poor, isolated the minorities, and was a profound contributor to the crisis which is now taking place in New York City. We are seeing the greatest city in the United States deteriorating and dying—in considerable measure because of the unintended and unplanned consequences of federal social investments in transportation, in housing, and in agriculture. We paid billions upon billions of dollars for the crisis in the United States that we are suffering today.

I could take that point and apply it to practically every department of government. But it is not because Eisenhower was an evil man, or because the Congress of the United States that went along with him was composed of evil men and women. We have a society in which one out of six workers is in some way involved in the automobile industry and in which three or four of the top ten major corporations are oil corporations. In such a society, it is just plain common sense to maximize the priorities of auto corporations and petroleum corporations rather than of working people, cities, minorities, women, and so on. This is built into the society; it is not conspiratorial.

Nonetheless, of course, there are conspiracies. The *Wall Street Journal*, which has a penchant for the idealization of greed and nastiness that never ceases to surprise me, came out with the following program for President Jimmy Carter: Cut down on social spending and increase tax exemptions for the corporate rich. The reason is that there is a capital shortage which the corporate rich discovered about two years ago. Nobody else has seen it, but Secretary Simon, President Ford, and the *Wall Street Journal* have been motivating programs for about two years on the basis of a nonexistent capital shortage. We are absolutely awash in liquidity in this country and the government is being asked to create more liquidity by giving privileges to the rich. There is that kind of consicous attempt to manipulate the system on behalf of the rich. But I will say that the *Wall Street Journal* probably believes that this policy would create the greatest happiness for the greatest number of people.

In conclusion, the issue before American society is not whether there will be national economic planning. We already have it. Our problem is that the plan on which we operate today proceeds according to the priorities of a hidden agenda. That hidden agenda, which has a remarkable consistency, is that government shall on all major decisions maximize corporate priorities. I suggest therefore that the issue is whether the plan will become democratic with a small "d," transparent and open, or whether it will remain bureaucratic and secret.

And, in terms of that point, one last unkind word for the American corporation. I believe the American corporation is now becoming decadent. The typical corporate executive, after reading the *Wall Street Journal* editorial in the morning about the glories of free enterprise and the free market system, immediately tries to figure how to get another subsidy from the government. There is a sense—and please understand me clearly, because there are enormous differences between the

United States and the Soviet Union—but there is a sense in which executives in the United States are increasingly playing the role that commissars play in the Soviet Union. They act as a bureaucratic elite with a government at their disposal, taking money from the people to fulfill their priorities and their interests and to make their decisions.

My second point, then, is that capitalism, which was such a magnificent and revolutionary system, has by its very socialization now created a corporate collectivist society in which government and the corporations cooperate with the public's money on the basis of corporate priorities.

III. THE ALTERNATIVES TO CAPITALISM

What are the alternatives? What would be a perspective for freedom and social justice in the modern world? Now let me state at the outset that I am a socialist, a democratic socialist. That is, I do not regard any of the totalitarian societies which claim to be socialist as being socialist. I think they are bureaucratic collectivisms, not democratic and free collectivisms. But much of what I am going to say does not require that you be a socialist to agree with me. I hope everybody becomes a socialist. However, I would like to illuminate the minds of those who perversely will not go the whole way but are only willing to accept this or that reform. Many of the things that I am about to say are acceptable to any intelligent, humane liberal who does not want to transform the basic system but to make it better.

The principle of the alternative to corporate collectivism, it seems to me, is the democratization of economic and social decision making. This is the fundamental principle that one then wants to follow in absolutely every aspect of the society. Under capitalism, by an (I hope) imperishable conquest of the human spirit, the idea of political freedom came into being, and we are eternally in the debt of the capitalist society for that. But now the concept of freedom and democracy must be applied to areas in which capitalism never applied it. Unless we democratize our increasingly collectivist state, which acts on the basis of a hidden agenda, we will lose freedom. Therefore, I believe that the democratization of economic and social power is the fundamental principle.

Now let me suggest some very specific ways that this can be done. First, there is a very moderate bill to which the Democratic Party platform was formally committed in 1976 and which a good many Democrats are busily running away from as fast as they possibly can.

The bill is the Hawkins-Humphrey Full Employment Bill. It is a bill for the most conservative modicum of liberal capitalist planning. It says that the President of the United States[3] every year will make an analysis of all the investment decisions, public and private, that can be anticipated for that year, add up those investment decisions, and discover what level of unemployment will result, given the population, technology, and so on. If adult unemployment exceeds 3 percent, which in plainer English is about 4 percent total unemployment, then the President is required by law to present to the Congress in an annual message those programs which will make up the shortfall and reduce unemployment to 4 percent or 3 percent adult unemployment. It is a very modest bill. It is viewed in some quarters as being practically the equivalent of Bolshevism. But it would begin to attack the most outrageous and abiding evil in this society today, its high level of unemployment, which is not randomly distributed but particularly discriminates against minorities, youth, and women.

This would commit us to reduce unemployment to 4 percent over the next four years. Sweden last year had 1.4 percent unemployment. We have not had anything that good since World War II. The problem is that many people are now saying we cannot do that, and some of them are liberals. Charles Schultze of the Brookings Institution says, in effect, that American society requires 5 percent unemployment in order to function. He maintains that as soon as we get to 5.5 percent unemployment, the inflationary tendencies become so strong that we must give up our campaign against unemployment in order to fight inflation. This implies that the people composing that reserve army of the unemployed, as Marx would describe it—this reserve army which is disproportionately black, Hispanic, female, and young—will pay for the struggle against inflation.

Now Schultze is absolutely right. If we accept American society today as an eternal given, we cannot achieve full employment without ruinous inflation. But the society is not an eternal given. One could, through the tax laws, redistribute wealth and have the corporate rich pay rather than the minorities, the females, and the young. One could have a redistribution income policy in the United States to deal with these problems. But this is the kind of issue that is not on the agenda. Here is a planning issue, a modest planning issue. A first tentative step is to have democratic decision making in the area of the economy. The achievement of this goal is going to take a tremendous struggle in this society, a struggle that is going to pit some of us against people who are Keynesian liberals but who say we need high unemployment to avoid inflation.

I view this Humphrey-Hawkins Bill, by the way, as only a first step. I think that we need a yearly national debate over the future. For example, the Auto Workers have raised a very interesting, profound question for the American society. In their 1976 contract negotiations, the UAW raised the question of the four-day week. I think we ought to have a debate in this society over something as important as that. Do we want to move from the five-day week to the four-day week as, among other things, a way of dealing with unemployment and environmental problems? It would mean we would take our increased productivity not totally in more commodities and goods, but in part in more leisure, which has its own problems but is less of a pollutant. We should have debates over these massive options. What we do in the society now is to back into the options on the basis of corporate priorities without ever really facing the fundamental issues.

Another proposal, an exceedingly moderate proposal, is endorsed not simply by Ralph Nader but by George Cabot Lodge of the Harvard Business School. George Cabot Lodge comes from the family who talks to God. Lodge and Nader and a lot of other good people have proposed that we have public and employee representatives on the boards of directors of every major corporation in the United States. How long can we kid ourselves that there is a managerial prerogative of secrecy for a company which has as much influence over our lives as IBM or U.S. Steel or General Motors? If people in the United States could elect directors to the boards of the Fortune 500 industrial and financial corporations, it would be more significant than most of the votes cast for members of state legislatures because these corporations are more significant for your life. Why do we say that the introduction of new technology, the location of plants, policies toward women, policies toward minorities, are private matters? Why is it in the United States today that we give corporations that leave New York City a federal subsidy by investment tax credit for leaving? What we need is to democratize and transform the corporation itself.

Related to this, I believe we should aim to democratize work. Most people in this society do not like what they do most of the time. That is a terrible commentary on the society. If you ask most Americans, "Do you like your job?" they will answer yes because they understand that the question you are asking them is, "Do you like your job compared to having no job?" Surely they do. If you ask them the question the Department of Health, Education and Welfare asked a couple of years back, "If you had to do it over again, would you take that job?" you find that most people in routine jobs, and particularly most blue collar people, do not like what they are doing.

Why do we not begin to change the way we work, for example, to allow workers to decide how to carry out a production process? One of the terrible things in American industry is that there are all kinds of workers who know how to do things better than the company does, but they have to hide it. If an automobile worker figures out a way to do something that is smarter than the way the company has figured it out, and he is quiet about it, he can bank two minutes of time on the production line and smoke a cigarette. But the foreman is walking around trying to figure out whether somebody has doped out a way to get two minutes from the company. So we lock up the productivity of our workers because they are caught in an antagonistic system in which they have to be crazy to give up their productivity so it can be expropriated by their boss. If we could begin to democratize the corporation, democratize work, democratize the economics of the society, perhaps we could get the benefit of some of that creativity.

As another point, I think we should have a Bureaucratic Relations Act in America, on the model of the National Labor Relations Act. The UAW has developed a very interesting institution called the Public Review Board. If a member of the UAW feels that somebody, some officer, some official of that union, has done him wrong, he has the right to appeal to an outside board not under the control of the union—a board which has been in existence now since the mid-1950's and has on innumerable occasions overruled the president of the union and the international executive board. Every bureaucratic institution of a certain size should be required to have a public review board. I think every university should have one. I think every large corporation should be forced to have one so that whenever the citizen is put into the situation where there is bureaucracy, he has the right of appeal. The bureaucrats would not decide the appeals against themselves, which is what happens in most of American society. These are just a few ideas and examples, not anything like a comprehensive list. I am suggesting that we should democratize the society from top to bottom in all of its institutions. That is the alternative.

I admit that I have omitted an enormous dimension of the problem. I have omitted it because it opens up so many difficulties that could not possibly be dealt with in the space allowed. I refer to the fact that there is a world outside of the United States. I do not believe that it is going to be tolerable much longer for 6 percent of the people of the world to consume 40 percent of the resources. But I leave that aside, precisely because it is so important that I cannot take it up at any kind of length here.

In conclusion, I suggest three propositions. First, capitalism was a magnificent advance for mankind, but a contradictory advance because it was a system of private socialization. The socialization brought enormous successes, but the private socialization has brought us intolerable contradictions. Second, the contemporary manifestation of the contradictions of private socialization are seen in an American system that is not a free enterprise system but a corporate collectivist system in which government honestly, sincerely, and non-conspiratorially follows corporate priorities because they are natural in this society. Third, the alternative, the way to pursue freedom in the late twentieth century, is to democratize economic and social as well as political institutions.

If we ask: "Shall we have free enterprise or shall we have collectivism?" we have asked a meaningless question, because there is no free enterprise today. There can be no free enterprise given the scale of our technology, given the interdependence of our world. Richard Nixon planned as much as Lyndon Johnson. If Ronald Reagan had been elected President of the United States, he would have planned as much as Lyndon Johnson or anybody else. It is not that people sell out. This is the basic thrust of our society.

Therefore, I think the question, as is often the case, is at least as important as the answer. I would reformulate the question and say the true question before us today in terms of freedom and social justice is: "Will the collectivist economic and social and political structure, which is emerging right now in front of our very eyes, be bureaucratically run by a united front of corporations and government on the basis of a hidden agenda moving toward a post-capitalist bureaucratic collectivist society run by the grandchildren of the executives?" Or is it possible for the collectivism which is emerging here and now to be democratized, to be subjected to the will of the people instead of its dominating the people? It is that second possibility that I am for, that I pose in terms of ethics and freedom and social justice. I am not sure it can prevail. I am not sure it will prevail. The one thing that I am sure of is that it is the only way for freedom to survive in the modern world.

NOTES

1. In this area, I would recommend to you a marvelous book. Though I think it is one of the greatest books of the twentieth century, it has not gotten all the attention it deserves: Rudolf Hilferding's *Finance Capital.*

It is a book which an obscure Russian by the name of V. I. Lenin also found interesting.

2. John Hicks, *A Theory of Economic History,* Oxford University Press, 1969, p. 162.
3. It has one interesting syntactic feature: "The President of the United States, he or she." It is the first bill that I know of that makes the grammatical admission that it is possible for a woman to become President of the United States.

4

A CAPITALIST CONCEPTION OF JUSTICE*

IRVING KRISTOL

It is fashionable these days for social commentators to ask, "Is capitalism compatible with social justice?" I submit that the only appropriate answer is "No." Indeed, this is the only possible answer. The term "social justice" was invented in order *not* to be compatible with capitalism.

What is the difference between "social justice" and plain, unqualified "justice?" Why can't we ask, "Is capitalism compatible with justice?" We can, and were we to do so, we would then have to explore the idea of justice that is peculiar to the capitalist system, because capitalism certainly does have an idea of justice.

"Social justice," however, was invented and propagated by people who were not much interested in understanding capitalism. These were nineteenth-century critics of capitalism—liberals, radicals, socialists—who invented the term in order to insinuate into the argument a quite different conception of the good society from the one proposed by liberal capitalism. As it is used today, the term has an irredeemably egalitarian and authoritarian thrust. Since capitalism as a socioeconomic or political system is neither egalitarian nor authoritarian, it is in truth incompatible with "social justice."

Let us first address the issue of egalitarianism. In a liberal or democratic capitalist society there is, indeed, a connection between justice and equality. Equality before the law and equality of political rights are fundamental to a liberal capitalist system and, in historical fact, the ideological Founding Fathers of liberal capitalism all did believe in equality before the law and in some form of equality of political rights. The introduction of the term "social justice" represents an effort to

* This paper is based on the transcript of a talk given at a symposium held at the University of Kansas in November, 1976.

stretch the idea of justice that is compatible with capitalism to cover *economic* equality as well. Proponents of something called "social justice" would persuade us that economic equality is as much a right as are equality before the law and equality of political rights. As a matter of fact, these proponents move in an egalitarian direction so formidably that inevitably *all* differences are seen sooner or later to be unjust. Differences between men and women, differencs betwen parents and children, differences between human beings and animals—all of these, as we have seen in the last ten or fifteen years, become questionable and controversial.

A person who believes in "social justice" is an egalitarian. I do not say that he or she necessarily believes in perfect equality; I do not think anyone believes in perfect equality. But "social justice" advocates are terribly interested in far more equality than a capitalist system is likely to deliver. Capitalism delivers many good things but, on the whole, economic equality is not one of them. It has never pretended to deliver economic equality. Rather, capitalism has always stood for equality of economic opportunity, reasonably understood to mean the absence of official barriers to economic opportunity.

We are now in an egalitarian age when Harvard professors write books wondering whether there is a problem of "social justice" if some people are born of handsome parents and are therefore more attractive than others. This is seriously discussed in Cambridge and in other learned circles. Capitalism is not interested in that. Capitalism says there ought to be no *official* barriers to economic opportunity. If one is born of handsome or talented parents, if one inherits a musical skill, or mathematical skill, or whatever, that is simply good luck. No one can question the person's right to the fruits of such skills. Capitalism believes that, through equal opportunity, each individual will pursue his happiness as he defines it, and as far as his natural assets (plus luck, good or bad) will permit. In pursuit of that happiness everyone will, to use that familiar phrase of Adam Smith, "better his condition."

Thus, capitalism says that equal opportunity will result in everyone's bettering his or her condition. And it does. The history of the world over the past 200 years shows that capitalism did indeed permit and encourage ordinary men and women in the pursuit of their happiness to improve their condition. Even Marx did not deny this. We are not as poor as our grandparents. We are all better off because individuals in pursuit of happiness, and without barriers being put in their way, are very creative, innovative, and adept at finding ways for societies to be more productive, thereby creating more wealth in which everyone shares.

Now, although individuals do better their condition under capitalism, they do not better their condition equally. In the pursuit of happiness, some will be more successful than others. Some will end up with more than others. Everyone will end up with *somewhat* more than he had—everyone. But some people will end up with a lot more than they had and some with a little more than they had. Capitalism does not perceive this as a problem. It is assumed that since everyone gets more, everyone ought to be content. If some people get more than others, the reason is to be found in their differential contributions to the economy. In a capitalist system, where the market predominates in economic decision making, people who—in whatever way—make different productive inputs into the economy receive different rewards. If one's input into the economy is great, one receives a large reward; if one's input is small, one receives a modest reward. The determination of these rewards is by public preferences and public tastes as expressed in the market. If the public wants basketball players to make $400,000 a year, then those who are good at basketball can become very, very rich. If the public wants to purchase certain paintings for $1 million or $2 million, then certain artists can become very, very rich. On the other hand, croquet players, even brilliant croquet players, won't better their condition to the same degree. And those who have no particular skill had better be lucky.

Capitalism does recognize, incidentally, that luck is a very important factor affecting the distribution of rewards in a market system. (As a matter of fact, all systems recognize this point except modern utopian systems, such as socialism.) Obviously, in order for a person to collect rewards of whatever kind, he has to be alive. The fact that he is alive is a matter of luck. He could have been hit by a truck. He could have developed cancer at the age of twelve. It happens to people all the time. All sorts of terrible things could have happened to reduce his income potential. The fact that they did not is luck.

Some of us who move about in the business world have met people who turn out to be millionaires. We wonder why. They don't seem very bright; they don't seem to have any special skills. Why are they millionaires? The answer is that they were in the right place at the right time and had just enough wit, in the old-fashioned sense of the term, to take advantage of their situation. Economic input under capitalism is a very peculiar and mysterious thing, which is one of the reasons that critics of capitalism get so upset about it. One never knows who is going to become enormously successful because one never knows what the market is going to reward. Who would have guessed that a Mr. Ray Kroc, who just resigned at age seventy-five as chairman of the board of McDonald's, would be worth $200 million or more? From what? Making hamburgers? The answer is that he saw something that others did not see. He saw that the American public would be happy buying hamburgers of high, standardized quality, served in a clean, attractive place. It sounds absurd to put it that way, because it is so obvious and banal. That is all he saw, and he invented McDonald's. There are in this country hundreds and thousands of people who are millionaires simply because they saw something that others did not see. They are not very brilliant. They just happened to be lucky and have the initiative to capitalize on it, as Mr. Kroc just happened to walk into a hamburger stand and get the idea for McDonald's.

This is the way the system works. It rewards people in terms of their contribution to the economy as measured and defined by the marketplace—namely, in terms of the free preferences of individual men and women who have money in their pockets and are free to spend it or not on this, that, or the other as they please. Economic justice under capitalism means the differential reward to individuals is based on their productive input *to the economy.* I emphasize "to the economy" because input is measured by the marketplace.

Is it "just" that Mr. Ray Kroc should have made so much money by merely figuring out a new way of selling hamburgers? They are the same old hamburgers, just better made, better marketed. Is it fair? Capitalism says it is fair. He is selling a good product; people want it; it is fair. It is "just" that he has made so much money.

However, capitalism doesn't say only that. It also understands that it is an exaggeration to say that literally *everyone* betters his condition when rewards are based on productive input. There are some people who are really not capable of taking part in the race at all because of mental illness, physical illness, bad luck, and so on. Such persons are simply not able to take advantage of the opportunity that does exist.

Capitalism as originally conceived by Adam Smith was not nearly so heartless a system as it presented itself during the nineteenth century. Adam Smith didn't say that people who could make no productive input into the economy through no fault of their own should be permitted to starve to death. Though not a believer, he was enough of a Christian to know that such a conclusion was not consistent with the virtue of charity. He understood that such people had to be provided for. There has never been any question of that. Adam Smith wrote two books. The book that first made him famous was not *The Wealth of Nations* but *The Theory of Moral Sentiments,* in which he said that the highest human sentiment is sympathy—the sympathy that men and women have for one another as human beings. Although *The Wealth of Nations* is an analysis of an economic system based on self-interest, Adam Smith never believed for a moment that human beings were strictly economic men or women. It took some later generations of economists to come up with that idea. Adam Smith understood that people live in a society, not just in an economy, and that they feel a sense of social obligation to one another, as well as a sense of engaging in mutually satisfactory economic transactions.

In both these books, but especially in *The Theory of Moral Sentiments,* Adam Smith addressed himself to the question, "What do the rich do with their money once they get it?" His answer was that they reinvest some of it so that society as a whole will become wealthier and everyone will continue to be able to improve his or her condition to some degree. Also, however, the rich will engage in one of the great pleasures that wealth affords: the expression of sympathy for one's fellow human beings. Smith said that the people who have money can only consume so much. What are they going to do with the money aside from what they consume and reinvest? They will use it in such a way as to gain a good reputation among their fellow citizens. He said this will be the natural way for wealthy people to behave under capitalism. Perhaps he was thinking primarily of Scotsmen. Still, his perceptiveness is interesting. Although capitalism has long been accused of being an inhumane system, we forget that capitalism and humanitarianism entered the modern world together. Name a modern, humane movement—criminal reform, decent treatment of women, kindness to animals, etc. Where does it originate? They all came from the rising bourgeoisie at the end of the eighteenth century. They were all middle-class movements. The movements didn't begin with peasants or aristocrats. Peasants were always cruel to animals and aristocrats could not care less about animals, or about wives, for that matter. It was the bourgeoisie, the capitalist middle class, that said animals should be treated with consideration, that criminals should not be tortured, that prisons should be places of punishment, yes, but humane places of punishment. It was the generation that helped establish the capitalist idea and the capitalist way of thinking in the world that brought these movements to life. Incidentally, the anti-slavery movement was also founded by middle-class men and women who had a sense of social responsibility toward their fellow citizens.

So it is simply and wholly untrue that capitalism is a harsh, vindictive, soulless system. A man like Adam Smith would never have dreamed of recommending such a system. No, he recommended the economic relations which constitute the market system, the capitalist system, on the assumption that human beings would continue to recognize their social obligations to one another and act upon this recognition with some degree of consistency. Incidentally, he even seems to have believed in a progressive income tax.

However, something very peculiar happened after Adam Smith. Something very odd and very bad happened to the idea of capitalism and its reputation after the first generation of capitalism's intellectual Founding Fathers. The economics of capitalism became a "dismal science." One cannot read *The Wealth of Nations* and have any sense that economics is a dismal science. It is an inquiry into the causes of the wealth of nations that tells people how to get rich. It says, "If you organize your economic activities this way, everyone will get richer." There is nothing pessimistic about that, nothing dismal about that. It was an exhilarating message to the world.

Unfortunately, what gave capitalism a bad name in the early part of the nineteenth century was not the socialist's criticism of capitalism but, I fear, the work of the later capitalist economists. We do not even have a really good intellectual history of this episode because people who write histories of economic thought tend not to be interested in intellectual history, but in economics. For some reason, Malthus and then Ricardo decided that capitalist economics should not deal with the production of wealth but rather with its distribution. Adam Smith had said everyone could improve his condition. Malthus said the situation was hopeless, at least for the lower classes. If the lower classes improved their condition, he argued, they would start breeding like rabbits and shortly they would be right back where they started. Ricardo came along and said that the expanding population could not all be fed because there is a shortage of fertile land in the world. In his view, the condition of the working class over the long term was unimprovable.

This was the condition of capitalist economics for most of the nineteenth century. It is a most extraordinary and paradoxical episode in modern intellectual history. Throughout the nineteenth century, ordinary men and women, the masses, the working class, were clearly improving their condition. There is just no question that the working classes in England were better off in 1860 than they had been in 1810. In the United States there was never any such question and in France, too, it was quite clear that the system was working as Adam Smith had said it would. Yet all the economists of the School of Malthus and Ricardo kept saying, "It cannot happen. Sorry, people, but you're doomed to live in misery. There is nothing we can do about it. Just have fewer children and exercise continence." To which the people said, "Thank you very much. We do not much like this system you are recommending to us," as well they might not.

When the impossibility of helping the average man and woman through economic growth is rejected loudly and dogmatically by the leading economists of the day, many will believe it. When they conclude that their condition cannot be improved by economic growth, they will seek to improve it by redistribution, by taking it away from others who have more. It is nineteenth-century capitalist economic thought, with its incredible emphasis on the impossibility of improving the condition of the working class—even as the improvement was obviously taking place—that gave great popularity and plausibility to the socialist critique of capitalism and to the redistributionist impulse that began to emerge. This impulse, which is still so appealing, makes no sense. A nation can redistribute to its heart's content and it will not affect the average person one bit. There just isn't ever enough to redistribute. Nevertheless, once it became "clear" in the nineteenth century that there was no other way, redistribution became a very popular subject.

This absolutely insane folly on the part of capitalist economists, which gave capitalism a bad name, was followed by an even worse event. I do not like the term "free enterprise." I prefer the simple word, "capitalism." Here is why. The words "free enterprise" to describe capitalism, so far as I have been able to determine, enter our vocabulary in the latter part of the nineteenth century. "Capitalism" is a term invented by socialists, but it can be a neutral term, almost a technical term. There happened to be no other term around to describe that economic system, but various conservatively inclined, procapitalist people—procapitalist ideologues, I would say—decided they did not like the word "capitalism." They substituted "free enterprise" as part of a neo-Darwinian theory of the economy. "Free enterprise" emerged out of a conception of the economy and the society which said: "We'll all engage in competition, a war of all against all, and some of us will do better and some of us will survive; others will do worse and won't survive. That's tough and that's all there is to it. Of course, the majority will not survive." The majority did not greet this intelligence with glad tidings. The majority, on the whole, thought that this was not a very good way to live, that human society should not be constituted under the law of the jungle. So "free enterprise" and its connotations only continued to give capitalism a bad name.

I am convinced that capitalism has been slandered more by its defenders than by its critics. During the nineteenth century, as the socialists said that capitalism could not improve the condition of the working class, and capitalist economists agreed, the condition of the working class continued to improve. In the twentieth century, economists finally caught up with reality and decided that there really was such a thing as economic growth and that it was indeed possible, Malthus and Ricardo to the contrary notwithstanding. By then it was almost too late. For one century we had capitalism given a bad name by its friends and by its enemies, and it will take a long time to wash that reputation away, if it ever can be done.

Because capitalism after Adam Smith seemed to be associated with a hopeless view of the world, it provoked egalitarian impulses. Is it not a natural human sentiment to argue that, if we're all in a hopeless condition, we should be hopeless equally? Let us go down together. If that indeed is our condition, equality becomes a genuine virtue. Egalitarianism became such a plausible view of the world because capitalist apologists, for reasons which I do not understand, kept insisting that this is the nature of capitalism. Those who talk about "social justice" these days do not say that the income tax should be revised so that the rich people will get more, although there may be an economic case for it. (I am not saying there is, even though there might be.) "Social justice," the term, the idea, is intimately wedded to the notion of egalitarianism as a proper aim of social and economic policy, and capitalism is criticized as lacking in "social justice" because it does not achieve this equality. In fact, it does not; cannot, and never promised to achieve this result.

However, I think the more important thrust of the term "social justice" has to do with its authoritarian meaning rather than its egalitarian meaning. The term "social" prefixed before the word "justice" has a purpose and an effect which is to abolish the distinction between the public and the private sectors, a distinction which is absolutely crucial for a liberal society. It is the very definition of a liberal society that there be a public sector and a large, private sector where people can do what they want without government bothering them. What is a "social problem?" Is a social problem something that government can ignore? Would anyone say we have a social problem but it is not the business of government? Of course not.

The term "social justice" exists in order to identify those issues about which government should get active. A social problem is a problem that gives rise to a governmental policy, which is why people who believe in the expansion of the public sector are always inventing, discovering, or defining more and more social problems in our world. The world has not become any more problematic than it ever was. The proliferation of things called "social problems" arises out of an effort to get government more and more deeply involved in the lives of private citizens in an attempt to "cope with" or "solve" these "problems." Sometimes real problems are posed. Rarely are they followed by real solutions.

The idea of "social justice," however, assumes not only that government will intervene but that government will have, should have, and can have an authoritative knowledge as to what everyone merits or deserves in terms of the distribution of income and wealth. After all, if we do not like the inequality that results from the operation of the market, then who is going to make the decision as to the distribution of services and wealth? Some authority must be found to say so-and-so deserves more than so-and-so. Of course, the only possible such authority in the modern world is not the Church but the State. To the degree that one defines "social justice" as a kind of protest against the capitalist distribution of income, one proposes some other mechanism for the distribution of income. Government is the only other mechanism that can make the decisions as to who gets what, as to what he or she "deserves," for whatever reason.

The assumption that the government is able to make such decisions wisely, and therefore that government should make such decisions, violates the very premises of a liberal community. A liberal community exists on the premise that there is no such authority. If there were an authority which knew what everyone merited and could allocate it fairly, why would we need freedom? There would be no point in freedom. Let the authority do its work. Now, we have seen the experience of non-liberal societies, and not all of it is bad. I would not pretend that a liberal society is the only possible good society. If one likes the values of a particular non-liberal society, it may not be bad at all. There are many non-liberal societies I admire: monasteries are non-liberal societies, and I do not say they are bad societies. They are pretty good societies—but they are not liberal societies. The monk has no need for liberty if he believes there is someone else, his superior, who knows what is good for him and what reward he merits.

Once we assume that there is a superior authority who has authoritative knowledge of the common good and of the merits and demerits of every individual, the ground of a liberal society is swept away, because the very freedoms that subsist and thrive in a liberal society all assume that there is no such authoritative knowledge. Now, this assumption is not *necessarily* true. Maybe there is someone who really does have an authoritative knowledge of what is good for all of us and how much we all merit. We who choose a liberal society are skeptical as to the possibility. In any case, we think it is more likely that there will be ten people all claiming to have different versions of what is good for all of us and what we should all get, and therefore we choose to let the market settle it. It is an amicable way of not getting involved in endless philosophical or religious arguments about the nature of the true, the good, and the beautiful.

I emphasize that if we want a society that aims at social justice, then we must have a society with a powerful consensus on values, with no significant disagreements as to what is good, what is bad, what is desirable, what is undesirable, what the good life is, what the not-so-good life is. The community depicted in "Fiddler on the Roof" illustrates the problems that arise in non-liberal communities for those who do not accept community values. For those who disagreed with the community, who didn't believe in its kind of God, or who didn't believe in the community form of religious observance, there was only one thing to do: emigrate, leave. There was no liberty to live one's own kind of life in that community. I do not think it was a bad community. The values were respectable enough. It may even be regarded as a positively good community. On the other hand, it was not a liberal community.

Those who have an idea of "social justice" need an authority to enforce that idea of "social justice." What is the point of knowing "social justice" and not doing anything about it? Now, such enforcement is really self-enforcement in a monastery because this is a consensual community and everyone assents to its values. There are no big arguments taking place over values. But in our kind of society there are disagreements, and "social justice" can be achieved only if the government *imposes* consensus on people. In fact, this is what is happening to a greater degree every day. To impose "social justice," there must be an authority. That is, there must be someone who says he knows what it is. Those who make this claim can, further, claim power in the name of their knowledge of justice. Given power, they might possibly create a society that is better than a liberal society in some ways. However, I don't think this is likely in our complicated and diverse civilization. More likely, they will produce the kind of system found in the Communist nations or in other authoritarian nations where government insists that it knows what "social justice" is and imposes it on a dissenting people simply by force.

We cannot have a liberal society, with all the freedoms that prevail in a liberal society, if we glibly assume that "social justice" has a precise meaning which we can define and upon which we can agree, and which it then becomes the function of government to impose. In our world today, on the whole, churches no longer seriously claim an authoritative knowledge of the common good. In principle, the Catholic Church still does, but in fact it does not seem to do so actively. Most of the Protestant churches do not. The various sects of Judaism do not on the whole, except for the Orthodox, and they limit their authoritative claims to Jews. But there is a class of people who do believe that they can define "social justice," that they have an authoritative conception of the common good that should be imposed on the society by using the force of government. These people, products of our graduate and professional schools, for the most part, can be called "the new class."

Our society has produced a great many people who do not like the free market because they do not like a liberal society. They do not only dislike the fact that it has so much inequality in it. They also do not like the fact that it is so "disorderly"—which is why they are always so enamored of planning. But the notion of having a free society which is not to some degree disorderly is absurd. If there were perfect order, there could be no need for freedom.

There are a large number of people in the media, government, and the non-profit sector who are persuaded that they know better than the market how things should be distributed in our society. This group does not want people to be recompensed on the basis of contributions to the *economy*, which is the capitalist definition of justice, but on the basis of their contributions to the *society*—as the group determines the worth of that contribution. In a way, the perfect exemplars of this new class of people are persons like Ralph Nader and the graduates of Ivy League law schools. Having been born to middle-class, even affluent backgrounds on the whole, they are no longer interested in money, but are only interested in power, power to improve the world, power to make this a better place. How are they going to exercise this power? Through government, not by persuading you or me. They are not going to go out and sermonize; they are not going to lecture; they are not going to try to persuade. They are going to use the authority of government to achieve "social justice." I am not saying they will not achieve "social justice." I do not think for a moment they will, but I can't prove they will not. All I can say is that in their efforts to do so, they will destroy this liberal society. Those who have any affection for and attachment to this liberal society, as I do, will not view benignly these efforts at its destruction. One will take offense at the notion of transforming this into a planned society. Who is going to do the planning? Planning will be done by people who know what "social justice" is;

they will know exactly what investment decisions should be made, and to what end.

John Kenneth Galbraith provides an insight into the mentality of this group of "idealists." One theme running throughout Galbraith's books is his horror at the proliferation of toothpaste in the United States. He thinks it is terrible that we should walk into a drugstore or a supermarket and find so many brands of toothpaste. Think of the waste in advertising all these different kinds of toothpaste, which are not essentially different from one another! In a well-ordered, planned society, of the kind that Galbraith would like, there would be one kind of toothpaste—the best kind of toothpaste—and he, or people selected by him or his friends, would know what that toothpaste is. I myself would not view it as a tragedy if we only had one kind of toothpaste, but I think it would be a step in a certain direction. The fact that people buy all these brands of toothpaste tells me that, although John Kenneth Galbraith or even I may not see any difference in them, the consumers do. It is a harmless pleasure, it seems to me, to use the particular kind of toothpaste that one prefers.

In a planned economy, the consumer would not have that variety of toothpaste, that variety of all sorts of things, because that variety is a function of the market. It is only the market that panders to people. I use the word "pander" because I don't want to say that the market does only good things. The market responds to our weaknesses and our vices as well as to our virtues. It is the nature of a free society that it does so. To argue the case for "social justice" as is done today is to argue against a liberal society and a liberal economy in which the decisions about rewards are basically made by the market with, however, provisions for government intervention in cases of those who are in no position to compete in the race for rewards in the marketplace.

The notion of a "just society" existing on earth is a fantasy, a utopian fantasy. That is not what life on earth is like. The reason is that the world is full of other people who are different from you and me, alas, and we have to live with them. If they were all like us, we would live fine; but they are not all like us, and the point of a liberal society and of a market economy is to accept this difference and say, "Okay, you be you and I'll be I. We'll disagree, but we'll do business together. We'll mutually profit from doing business together, and we'll live not necessarily in friendship but at least in civility with one another."

I am not saying that capitalism is a just society. I am saying that there is a capitalist conception of justice which is a workable conception of justice. Anyone who promises you a just society on this earth is a fraud and a charlatan. I believe that this is not the nature of human destiny. It would mean that we all would be happy. Life is not like that. Life is doomed not to be like that. But if you do not accept this view, and if you really think that life can indeed be radically different from what it is, if you really believe that justice can prevail on earth, then you are likely to start taking phrases like "social justice" very seriously and to think that the function of politics is to rid the world of its evils: to abolish war, to abolish poverty, to abolish discrimination, to abolish envy, to abolish, abolish, abolish. We are not going to abolish any of those things. If we push them out one window, they will come in through another window in some unforeseen form. The reforms of today give rise to the evils of tomorrow. That is the history of the human race.

If one can be somewhat stoical about this circumstance, the basic precondition of social life, capitalism becomes much more tolerable. However, if one is not stoical about it, if one demands more of life than life can give, then capitalism is certainly the wrong system because capitalism does not promise that much and does not give you that much. All it gives is a greater abundance of material goods and a great deal of freedom to cope with the problems of the human condition on your own.

5

FREE ENTERPRISE AS THE EMBODIMENT OF JUSTICE

JOHN HOSPERS

I

The above title contains two terms which require clarification: "free enterprise" and "justice." The first is comparatively easy; the second a little more difficult.

The free-enterprise system is an economic system in which all production and exchange of goods and services takes place without interference by the state. Whenever government intervenes in this system of voluntary exchanges, we have what is called "interventionism," and to the extent that interventionism exists, free enterprise does not. Let me make one qualification at the outset: I shall not consider it inconsistent with the free-enterprise system for the state to intervene in human affairs in order to keep the peace: the police, to apprehend those suspected with good reason of committing crimes (real crimes, not "victimless crimes"); the armed forces, to protect the inhabitants against attack from foreign nations; and the courts, to adjudicate disputes. These are indeed interventions by the state into human affairs, but for the sake of preserving or restoring peace. Economic activity does not flourish against a background of civil disorder or the rubble of war, and thus, such intervention is to the benefit of economic activity, not to its detriment. Thus, by "free enterprise" I shall mean the absence of state intervention in the mechanism of the free market, in the production and distribution of goods and services, other than for fulfilling the peace-keeping functions just described.

When I talk about justice in this essay, I shall not be talking about justice as dispensed by judges or juries or policemen or parole boards. My topic here is specifically justice in connection with the free-enterprise system, and this being an economic system, my sole concern will be with that.

In 1976 we celebrated not one but two bicentennials: one, the political bicentennial of the Declaration of Independence, with which we are all familiar; the other, which to my mind is equally important, the publication in 1776 of Adam Smith's great classic *The Wealth of Nations.* This book set forth systematically the principles underlying the free-enterprise system. The first event of 1776 placed severe limitations on the powers of the state, attempting to make certain forever that it was to be the servant and not the master, and the eternal guardian of the rights of man. The second event contained a formula for growth and prosperity without interference by the state.[1]

The Founding Fathers of our country, when they created the political structure which alone could guarantee enduring prosperity, had no idea that this is what they were creating; they thought only that they were creating a bulwark against oppression and freedom to worship as they chose, freedom to express opinions, freedom against tyranny by the state. They had no idea of an industrialized technological society in which a high standard of living would prevail and drudgery be reduced to a small fraction of what it had been before. They would not have denied economic freedom, of course; they would have insisted on freedom of production and trade, uninterfered with by the state; but they had no idea what a fantastic edifice could and would be built on that political, or anti-political, foundation. Certainly it was one of the most fortunate marriages to occur in human history, and the offspring of that marriage was the United States of America.

Of all the institutions created in America, the one that has contributed most to the well-being of its inhabitants, more even than any of its institutions of government, is the free-enterprise system. Its beneficiaries are largely ignorant of it; many people, who are its beneficiaries, rail against it; yet it has provided more people with a high standard of living, by far, than any other system ever devised by man. The most ordinary things, which even the poorest people in America take for granted, were either non-existent or extreme luxuries in pre-industrial times. Because of free enterprise, and the technology which it developed, many times more people can live on the face of the earth than had ever been dreamed possible before. America was already over-populated with half a million Indians, considering the large amount of space each one needed to survive on hunting and fishing, whereas today we support over 200 million in comparative affluence.

What is the secret of this system? In a word, it is liberty: liberty to think, to invent, to produce, to trade. When no restrictions are placed on these things, the ingenuity of man knows no bounds. But almost always in past history, such restrictions have been there: kings, emperors, tribal chieftains, modern governments drain off so much of the energies of the people, especially the creative people, whom they perceive as threats to their power, that the energies of men are sapped and their spirits broken. When America began, living conditions had hardly changed since ancient times. Even in Western Europe until 150 years ago, people lived mostly in windowless hovels with a hole in the ceiling for the smoke.[2] In America, men were free to produce and trade as they chose. There was a tremendous outburst of outgoing energy and upward mobility; rags-to-riches stories became daily occurrences. There were no hereditary castes, no strangling bureaucracies, no heavy hand of government to keep them down. And thus, already by 1825, the United States was the most prosperous nation in the history of the world. In one century, three generations, this release of human energy had created an entirely new world.

Liberty bears the same fruits today. Consider East Germany versus West Germany: people of the same culture, the same traditions, the same level of technical skill, torn asunder by the accidents of war. On the one side, tyranny, regulation, misery, and an irresistible impulse to leave; on the other side, capitalism, freedom, and affluence. In the Far East, Malaysia, Taiwan, Hong Kong, Japan all rely primarily on free markets; all are thriving, their people full of vital energy—a far cry from the socialist states, India, China, now Vietnam and Cambodia, all relying on central planning and heavy with the dead weight of bureaucracy.

But the contrast to which I want most to call your attention is that of Japan versus India. India has many more natural resources than Japan; the British in 1948 left India an excellent railroad system, factories, equipment, functioning political institutions, skilled administrators, men trained in industrial techniques. Top Indian workers were as efficient as any in the world, as they still are outside India. Japan had no comparable advantages; she had to provide all her capital from domestic sources after the Meiji Restoration of 1867. Yet Japan became prosperous, and at its present rate it will soon be more prosperous than the U.S. Why the difference?

Japan followed a free market system, taking Great Britain of its time as its model. It never tried to control the amount or direction of investment or the structure of output. But Indian leaders (of 1948) were schooled in Fabian socialism and central planning; they took Russia as their model, and a series of five-year plans with detailed programs of investment by government in whatever enterprises its political leaders chose.[3]

II

Yes, but does the free-enterprise system provide justice? Well, "justice" is a big word, and there is endless dispute about what it even means. Let me take its most precise meaning in ordinary usage, that of treating each person in accordance with his or her deserts. Does the free-enterprise system do that?

Well, I can't say that I know what each and every person deserves—nor that anyone else does, either. The word "deserve" itself has come in for much questioning—and quite deservedly so. If I were to embark on an analysis of such concepts as justice and deserts here, it would take me far afield from the topic of free enterprise and justice. I shall therefore take a somewhat different approach: I shall leave undisturbed whatever conception of justice you already have, and then advance one consideration after another to see whether it accords with justice as you conceive it.

1. Certainly the free-enterprise system has provided affluence to an extent undreamed of in history, extending even to the least well-off segments of society. Do all the members of society deserve this affluence? I for one like to see people as well off as possible, and as free from drudgery and dawn-to-dusk labor, and I am happy that the free-enterprise system has made this possible. Most of those who are the beneficiaries of this legacy have done little to deserve it: the medieval serf, working with his unaided muscles, worked much harder and had much less to show for it than even the lowliest worker in America today. And in this respect, the American is just lucky; he lives in a post-industrial age, and in a country where economic liberty is permitted. Left to his own devices, he would be unable to invent even one of the labor-saving devices which have so reduced drudgery in Western civilization. Most people rely for their affluence on the inventive genius and hard work of a comparatively small number of people, whom they could not repay in a thousand lifetimes for the benefits that this small group has conferred upon them. But even though most of mankind's benefactors are lost to history, unappreciated and unknown, I am glad for the windfall that has been conferred on the rest of us as a result. Even today, I can never pick up a telephone and dial a city 3,000 miles away without a feeling of gratitude to all those who made it possible. I know that *I* couldn't have done it; I am only grateful that *they* did.

Every step in the long march of mankind toward affluence has been an improvement for all future generations to enjoy; every invention that makes possible more production per man-hour of work is a benefit to generations yet unborn at the time the invention was made. It is always the substitution of tools and machines for the unaided muscles of human beings. As a result of this, people today do far less work, and work fewer hours, than did any of their ancestors. Many workers whose ancestors toiled sixteen hours a day, even so barely keeping body and soul together, today work eight hours or less pushing a few buttons on machines. Even the worker who does just about the same thing as his ancestors did, like the town barber, benefits: not only in his possessions but in his wages. His ancestors received far less for doing the same thing that he is doing; and this is because in the general rise in standard of living, owing to free enterprise, he is carried along on the rising wave: unless he earned a comparable wage, he would not go into this trade or profession. If justice implies that a man receive as much as possible for his effort, then the free-enterprise system is the only one that even remotely approaches justice.

The free-enterprise system cannot prevent natural catastrophes, such as hurricanes and earthquakes. A farmer's crops may be ruined by drought; and if you ask whether this is just, whether the farmer deserves this fate, I would say no—although some would restrict talk about justice and injustice to the relations of human beings to one another, and in this case the drought would be a misfortune but not an injustice. In any event, no economic system can prevent certain ravages of nature; to some extent technology can, and if the economy is left free, it will.

2. The free market ". . . is, in fact, an enormous computer, far superior to any electronic computer man has ever devised . . ."[4] It responds to shortages by increases in prices. But it spurs others to enter that area because of the lure of profit, and thus benefits the consumer as the price comes down when supply increases. It functions automatically and efficiently in a way government intervention and planning can never match.

3. In the free-enterprise system, wages are determined by the free market. Is the free-market wage a just wage? I could, of course, settle the matter definitionally (as some writers have done) by simply defining "the wage you deserve" as "the wage you get on the free market." I shall not do this, however, since it begs the question at issue, though it does point to something true and important: the wages one earns are not determined by the dictum of some official or committee, but by impersonal forces of the market. Not only is this preferable to the alternative, but it corresponds more closely to our idea of justice: if your job is dirty or dangerous, we would ordinarily say you deserve a higher

wage for performing it; and since people will not go into that line of work without additional incentive, the wages (on the free market) are indeed higher for such work. Others, such as physicians, who require extensive and expensive training will not go into it unless they can command greater income after they have achieved it, and of course, physicians on the free market will earn more than average income. (If you think that some of them get *more* than they deserve, and I agree, remember that part of this is not the result of the free market but of arbitrary restriction on the number of physicians by the A.M.A., through which the government licenses physicians; the fault here is with government licensure, not with the free market.)

4. In general, the free market rewards effort and initiative. It makes it worthwhile for people to invent and produce; it rewards the Edisons of the world for their efforts, though not always immediately. Some innovators, to be sure, die unrewarded for their efforts, only to be recognized by posterity, but I see no way around this; this would be even more the case in a bureaucratic Soviet-style system in which a government official or committee of officials determine who should be rewarded (the committee is much more likely to be mistaken). In the U.S.S.R., in one five-year plan, physicians get higher wages than teachers, and in another plan they don't; in free enterprise, at least the physician is more likely to have income commensurate with his actual achievement.

5. In a free-enterprise system a person who decides to spend money on a new venture does so at *his own* risk. He makes the decisions, he gets the profit or bankruptcy; he doesn't force *us* to share the consequences of *his* decisions. But when the state goes in for new ventures, it risks your money and my money on things and projects we may not even approve of, or see to be foolish from the beginning. To be forced to pay for something you would never spend it on voluntarily, and believe to be useless, wasteful, or immoral, is hardly a model of justice.

6. And how, in any society, is a person to be rewarded in accord with merit? Merit by what standard? And who is to determine this merit? In a society in which the rewards are dispensed by a government bureaucracy, it is the bureaucracy, not the market, that decides, e.g., that engineers are more deserving than physicians and thus pays them more. What if someone believes that physicans are the more deserving? Or that income should depend on the individual case, not the category of labor involved? In a Soviet society, such a person is out in the cold. Someone else's idea of merit has won out. What you get depends on which group happens to be in power. This is surely a poor excuse for justice on practically anybody's definition, unless one holds

that bureaucrats are all-wise. In a free-enterprise system, at least, whether you succeed does not depend on some politician's decision as to your merit.[5]

In a free-enterprise society, one in which the state stays out of economic matters entirely (and I hasten to add that the U.S. does *not* have a free-enterprise society today), a person's financial success is determined not by his merit, certainly not by his moral merit, but by his *value* to others in the society, as they themselves estimate that value. A manufacturer who supplies a higher quality product at a lower price will usually capture a larger share of the market—that is, his product has a high value in the minds of consumers, and they respond by buying it in larger quantities. We may not care for the kind of product he sells, and we then refrain from buying it. But the marketplace is a kind of economic democracy in which every time a person buys a product he casts a vote for that product, and a part of his reward goes to the manufacturing company that produces it, often enabling it to mass-produce it in larger quantity and lower the price still further, as well as adding more employees. It may be in our eyes a perfectly useless product, some knicknack which we believe people would be better off without. But if they nevertheless buy it in quantity, this shows that it has a value *to them;* they willingly buy it in preference to something else. No government official decides how much reward the manufacturer gets; the market, as expressed in the collective decision of countless consumers, decides. And if the manufacturer is not subsidized by government, as in a free-enterprise system of course he is not, we do not pay a penny toward the production of any product we do not want; if you do not want it, we are free not to buy it.

The amount of financial reward that goes to the producer of the product (owner, manager, workers) depends not on the public's evaluation of his moral desert but on how valuable his product is *to them.* If another manufacturer worked twice as hard and succeeded only in producing an inferior product, we do not feel that we owe him as much for his effort as we do to the first producers, whose product is superior but whose effort may have been less. As Hayek puts it, the determining factor "is the advantage we derive from what others offer us, not their merit in providing it. . . . The mark of the free man is to be dependent for his livelihood not on other persons' views of his merit but solely on what he has to offer them."[6] And no one can make a million dollars on the free market unless he offers them a product they are willing to buy in sufficient quantity to return that million dollars. Unlike the State, the president of General Motors has no powers of extortion, arrest, or trial.

All this, thinks Hayek (and I agree with him), is as it should be. "It is more than doubtful whether even a fairly successful attempt to make rewards correspond to merit would produce a more attractive or even a more tolerable social order."[7] We cannot have government in the business of apportioning reward to desert without the government becoming a vast and omnipotent busybody, with a huge investigative and enforcement squad (super-duper snoopers) prying into the details of everyone's life and leaving them no privacy. The cure is much worse than the disease; as the British philosopher F. R. Lucas put it, "Better have wealthy plutocrats than omnipotent bureaucrats."

III

Let us now examine some of the things that the state does, in the charred and battered remains of the free-enterprise system which is the United States today, to see whether *they* accord with justice.

How much money does the government spend? According to a recent monthly bulletin of the National Taxpayers' Union, *Dollars and Sense,* entitled "Taxpayer's Liability Index," the following is a current "statement of account" for each man, woman and child in the United States:

National debt: $701 billion. Your share, $14,020
Accounts payable: $81 billion. Your share, $1,620
Undelivered orders: $266 billion. Your share, $5,320
Long-term contracts: $13 billion. Your share, $260
Government guarantees: $190 billion. Your share $3,800
Insurance commitments: $1,629 billion. Your share, $32,580
Annuity programs (including social security): $4,650 billion. Your share, $93,000
Unadjudicated claims and international commitments: $53 billion. Your share, $1,060
Total, $7,583 billion. Your share, $151,660
Terms: you just keep paying.

Most Americans have very little idea what their government is doing to them. They do complain often about high taxes and high prices, and some complain about government regulation; but very few have any idea of how they have been victimized.

* In 1900 the United States had 76 million people. The total expenditure of the federal government was $525 million. This is a per capita cost of $6.90—what it costs per person per year to run the federal government.

* In the 1978 federal budget, planned federal expenditures are over $500 billion—over $2,500 for every man, woman, and child in the United States.

* This is over $10,000 for the average family of four. This amount is not all raised out of taxes—more of this shortly. But it is what the federal government is *spending* out of the income of the family of four. (In 1940 the figure was $308.)

* It took sixty years, from 1789 to 1848, for the U.S. government to spend one billion dollars. In 1976, it spent that amount in two-thirds of one day. (It spends $1.5 billion more per week than it collects.)

* The total spending in the fiscal year ended July 1975 for all levels of government (federal, state, local) was a little over $555 billion—twice the amount it was eight years ago, over twenty-five times the amount in 1940; about $2,750 for every man, woman, and child in America. If the present rate continues, it will double again in the next six years.

If we consider not only the current expenditures but what our government has committed us to for the future, the figure is much higher. Your share of the national debt alone is $14,020—that is the amount to which the federal government has committed you (and the same for me, and the same for every other American), presumably to be paid by our children and grandchildren; we are all in hock to that amount to sustain the national debt alone.

But even that amount is small compared with others. The estimated future claims on the Social Security System by the people now paying into it is, on a conservative estimate, $1.5 trillion (thousand billion)—enough to break the system by the mid-1980's, unless the payments taken out of each paycheck increase astronomically. The chances are that by the time you need to get back what you have been forced to pay in, the cupboard will be bare. And Social Security is only one of the government annuity commitments; together they total $4,650 billion ($93,000 for each American). The total advance commitments, to be paid out of future taxes, is a shade over seven and a half trillion dollars, $151,660 for each American alive today (up $13,000 from last year). While you have been working to make ends meet, the politicians you elected to office have been racking up debts on your behalf—often in the name of "social justice."

What is it spent on? The money you've already paid to sustain the government, and the much higher amount that it has you in hock for, has been and is being spent on an endless procession of projects—some of them merely wasteful, others positively crazy. Here are a few examples of thousands: $37,000 for a potato chip factory in Morocco (Morocco grows no potatoes); $32,500 to officials in Kenya to help them purchase extra wives; a million dollars for the Air Force to purchase a simple tent (which blew away after a few hours' service); $415,000 was spent to remodel a jet-set hotel in Haiti where the cheapest rooms cost $150 a day.

And so, the wild spending continues. The federal government has funded projects on hitchhiking, why people fall in love, primate teeth, African climate during the Ice Age, and the social behavior of Alaskan brown bears (*Av Reporter*, January 1976, p. 6).

But now, we get to bigger items. There was the $57,000 per *minute* to pay interest on the national debt. And ranging back over the last twenty years, there is a little item of $250 billion for foreign aid, which has enabled the U.S. to finance and support both sides in a dozen wars. I could go on with this indefinitely, but let me turn immediately to the biggest ripoff of them all: two years ago, the total cost of federal welfare projects (not including state and local ones), including relief payments, Model Cities projects, housing and urban renewal, and similar federal efforts at income redistribution, totaled more than $287 billion per year (it has gone up since)—that is, about $1,350 per year for each and every American alive.

In fact, the total cost of the federal government has doubled between 1964 and 1974. Of this increase, 19 percent has been for military spending, the other 81 percent on domestic welfare-type programs. Federal aid to education jumped 466 percent during this period; public assistance 235 percent; health care and medical services 4,571 percent—all for the federal government alone. Between 1930 and 1974, the population grew 71 percent and the federal bureaucracy 462 percent. About 80 percent of the cost of Lyndon Johnson's "Great Society" programs went to pay for the salaries and offices of the government employees who administered the programs.

When you spend the money you yourself have worked to earn, you are likely to exercise some care in how you spend it; you know that it will take lots of work to replace what you spend. If you do spend it foolishly, the consequences are on your own head, and you may learn your lesson and exercise some restraint next time. But when government takes it from you and spends what you might otherwise have spent as *you* saw fit, these restraints are totally abandoned. One almost never spends other people's money as carefully as one's own money.

In private enterprise, when you get too many men in the front office, the overhead becomes too great and you have to cut down expenditures to meet the competition. But in a government bureaucracy it does not work that way; you prove the importance of your department, not by trimming expenses and operating more efficiently, but by adding to your employees (whether you need them or not) and increasing the make-work or busy-work to give the impression that many important things are getting done. It is no accident that government is always boundlessly wasteful in its expenditures; this is not a mistake that can be "corrected the next time around." It is of the very nature and essence of governmental activities and the way they are funded and managed.

Where does the money come from? (1) Taxation. How are all these expenditures met? The largest part, though not all of it, comes from taxes.

Prior to the Civil War, when the federal government adhered to the Constitution and took on only such functions as were delegated to it, the entire income of the U.S. government came from excise taxes and tariffs. Property taxes were also levied by individual communities to meet their needs. Then came the federal income tax in 1913, which had previously been declared unconstitutional by the Supreme Court. At first, the tax was quite small; but the hole in the dike became larger each year, as politicians saw a chance to loot the people's earnings, promising them all kinds of goodies at public expense in order to get more votes. By 1929 the tax collector was taking $1 of every $10 earned. By 1930, it was $1 in every $5. Late in the 1950's it became $1 in every $4. Today federal taxes take approximately $1 in every $3 earned. Today the government takes in taxes the entire income of every American living west of the Mississippi River.

In addition to federal income tax, and numerous state and local taxes, and sales taxes on items bought, and property taxes, and inheritance taxes, there are numerous other "concealed taxes" which most people do not suspect even exist. There are, for example, the 150+ taxes on every loaf of bread, accounting for more than half the cost of the loaf of bread, even in those states where food is not subject to sales tax. There are taxes on the farmer who grew the wheat—his property taxes, and so on, which means he has to get more for his wheat in order to break even. There is the gasoline tax on the driver who drives the truckload of wheat to the mill, his driver's license, license plates, etc. The miller who grinds the wheat has to pay social security taxes, business licenses, property taxes, state board of equalization taxes, and countless others—and to break even, he has to charge more because of these costs imposed by the state. And so it goes: all those taxes have to be added to the cost of every loaf of bread you buy.

What is the result? As taxes mount, more and more small businesses, and some large ones, are squeezed out and go bankrupt. They are the victims of a government that is endlessly hungry for more of their earnings. When a government demands and gets a larger share of everyone's income, all of us are poorer. The government receives more and more, and each of us is left with less and less. The corner shopkeeper—already just breaking even, thanks to the government's tax bite—can no longer continue in business and is forced to close his doors and dismiss his employees. The same goes for the spare parts factory down the street, and the used-car garage, and the restaurant around the corner. (The vocal champions of "social justice" never seem to think of them.) In every city and town, thousands of people are laid off by companies that could not continue to see their life's savings and the effort of years go up in smoke because of the ever-increasing tribute levied upon them. The employees of these companies, of course, are out of jobs too, and those who still have jobs have increasing amounts deducted from their paychecks; they too have become tax-slaves of the government. Nor does it end there: customers patronize these factories and stores and restaurants less because they too are hit with added taxes. They cannot afford to go to the restaurant as often (so the restaurant closes its doors) or to do anything not absolutely necessary. And thus, the higher taxes and fewer customers hit the shopkeeper both at the same time. The government has tightened the vise on each one of them to such a point that they can no longer exist as parts of a functioning economy.

If the full story were known, people would rise in rebellion against this legalized looting. The trick is to divert the blame; and in this respect, the government propagandists have been fairly successful. When the price of groceries goes up, people picket the supermarkets. But the supermarket owner can't help the price rise. He would like to lower the price and undersell the competition, but he cannot, because (thanks to the increased government levy on *his* suppliers) the prices he is charged have gone up also, and he cannot operate for long at a loss. He has to increase his prices or go out of business. But so do his suppliers, who are also hit with increased costs of production, thanks to increasing demands from government. And so on.

Instead of picketing the supermarket, it would be more sensible to picket the U.S. Department of Agriculture buildings. The Agriculture Department subsidizes thousands of farmers to grow (or not to grow) crops, which increases their prices in the markets. You have to pay more for the groceries because of the shortage artificially created for what the government paid the farmer not to grow (or for what was grown and shipped to Russia, causing a shortage here).

(2) *Increasing the debt.* Even the crushing burden of taxation is not enough to fill the bottomless pit of government expenditure. As a stop-gap measure, the national debt gets increased at every session of Congress. This means that the American taxpayer is hit with another IOU; the services the money is used to give us are to be paid for later, out of our future earnings and those of our children and grandchildren.

There is a popular misconception that the national debt is "meaningless" because "after all, we only owe it to ourselves." Well, if it is no burden, why not increase it a tenfold, a hundredfold? But there is a catch in it. "We owe it to ourselves" means "Some of us (or our descendants) will have to pay for it, in order that those others of us who are on the receiving end of it may continue to receive it."

If it were to be paid next year, it would take more than the entire federal budget to do it. The idea was to pay it off gradually over the years, during years of prosperity when there would be a budget surplus. But even during prosperous years there was almost always a deficit, as politicians thought of new ways to spend our money. The chances of ever paying it off are more remote each year; meanwhile, American taxpayers are paying over $30 billion a year just to pay the *interest* on it. At the present rate, the annual interest on the debt will pass $100 billion by the end of the century. If every American who files an income tax return this year were to add an amount sufficient to retire the debt, it would require $6,500 from each taxpayer.

If the debt were repudiated, what would be the result? It would be like an individual declaring bankruptcy; all those to whom the debt is owed would be unable to recoup it. Everyone looking forward to Social Security checks, for example, would find himself out in the cold. For the government to repudiate the debt outright would be political suicide; what is more probable is that it will gradually get paid (in part) in depreciated currency—currency depreciated through inflation. The government will pay in printing-press money, which will gradually become next to worthless, but technically it will have been paid.

(3) *Inflation.* And so, we come to the most likely source: debasing the currency through inflation. In ancient times, when the emperor wanted to spend more but paper money was unknown, he would have the coins clipped at the edges and then declare by fiat that they were worth as much as before; thus, he would have gold and silver to melt into new coins. But with the invention of paper currency, it became much easier: just print pieces of paper (called Federal Reserve Notes), make them redeemable in gold or silver for as long as people have enough faith

in government to refrain from turning in the paper for metal, and then pull the rug out by declaring that the paper is no longer redeemable in gold or silver—in other words, that it has nothing to back it.

As government projects proliferate, they have to be financed somehow. When the people have been milked through taxation all that they can be without rebelling, and the debt increased, the next and easiest step is to print up more paper money (or, what amounts to the same thing, issue credit to the banks which they can use to make loans). Inflation is simply an increase in the supply of money and credit. Higher prices are not inflation but the *result* of inflation. Higher prices are inevitable because when the government prints more green paper, you have more money without an increase in goods; the money is worth less in relation to the goods, and the price of everything goes up. If everyone had tomorrow double the amount of paper money he has today, it would avail him nothing; he would have the illusion of increased wealth, until he went to the store to find that the price of everything had increased proportionately. If the U.S. Congress continues inflating at the present rate, a $50 bag of groceries will cost $100 in 1986, $200 in 1996, and $4,000 in 2006—and this is only if the rate of inflation continues at 6.8 percent; almost no one believes any longer that the rate will remain that small.[8]

To inflate the currency is a tempting and practically irresistible device of governments, to pay for the promises the politicians made to the voters ("Vote for me, and I'll give you so-and-so"). At first, the result has the illusion of success, because it takes time for the new currency to percolate through the economy, and the first persons to get the new money do not yet experience the rise in prices—so the first effect is that of increased prosperity. It is rather like a heroin shot: at first, there is the euphoria; but as that wears off, then comes the withdrawal, when people realize that they are paying much more for the same thing. Meanwhile, those on fixed incomes, and those whose income does not increase in proportion to the increase in the money supply, are caught in the squeeze; the government, by its inflationary devices, has robbed them as surely as if it had held them up at gunpoint. But by the time the withdrawal stage comes, there is probably another President in office to take the blame. Most Presidents do not want to be blamed for an inherited mess, and most of the voters know nothing of the causes of inflation, so those in charge of government do the easiest thing: they start another round of inflation, more paper, higher prices, until of course the next withdrawal inevitably comes. And each time, it takes more input to get less output (the law of diminishing returns). Thanks to inflation, the American dollar is worth less than 40 percent of what it was twenty-five years ago, and its value continues to decline with every passing year.

Regardless of their intentions, the men in government who do these things are thieves, stealing everyone's earnings and savings. Inflation is gradually making our savings worthless, making it no longer worthwhile to save, or even engage in productive activity except on a day-to-day survival basis. If you do manage to save something, taxes and inflation will make it nearly worthless anyway before the time comes for which you have saved it. But of course, it was all extracted from you in the name of "social justice."

Every earning American has already become a wage slave of the government. Americans who pay into the system through taxes are already less numerous than those who receive from the government in the form of relief benefits, food stamps, subsidies, veterans' benefits, wages as government employees, etc. The number of people who put the cookies into the federal cookie jar is decreasing in relation to the number of people who take cookies out of the jar. "Last year [1974] there was a total of 80,655,000 tax dependents: government employees, the military on active duty, the disabled and unemployed, retired and on welfare, vs. 71,650,000 non-government workers."[9] No civilization has ever crossed that watershed and survived.

There are many other insidious aspects of inflation. It drains your earnings of their value; it discourages thrift, savings, and industry. But the greatest danger of inflation is *loss of liberty.* This result is not as clearly or easily seen, but it is just as real. After inflation has raged for some time, dictatorship becomes almost inevitable. As inflation continues, public agitation for controls on prices grows. The government puts a ceiling on prices, and usually also on wages; and at that point the economy is in virtually total control of the government. None of this cures inflation as long as the government continues to print pieces of green paper. It only makes each dollar worth less, so that (with price controls) many products can no longer be produced at the required price except at a loss, and in the end they stop being produced. Thus, shortages develop throughout the economy; and in the wake of the shortages, riots and looting, for the absent products. Then a great outcry for "law and order" is heard, at almost any price. Out of this situation arises a Caesar—a dictator who restores law and order, unifies the people behind him, and takes their liberty from them in the name of security. This succession of events has occurred repeatedly throughout history.

Every totalitarian regime in modern times has ridden in on a wave of inflation and regulation: Napoleon, Mussolini, Hitler, Mao. The U.S. became the world monetary guardian after World War I and the collapse of the British pound; one shudders to think who may ride in on the collapse of the American dollar.

Some of us may be able to get through the period of galloping inflation; but who of us can protect himself against the dictatorship to follow? The millionaire on a lifeboat has no more than the pauper if the lifeboat sinks.

IV

To what end is all this money extracted from us by our government? Among other things, it is to *regulate* the very people who supply the government with the means to do it. The regulatory agencies, now so extensively proliferated, are so omnipotent in every sphere of economic activity that they threaten to strangle every industry they affect. I shall consider only a few examples of this regulation.

Regulation: (1) Agriculture. Most of the regulatory nightmares occurred after the government had begun to intervene on a large scale in the lives and businesses of American citizens—in which it had no constitutional right to interfere, but did so nonetheless and continues to this day. I mean the era ushered in by the New Deal in 1933, the watershed year which separates the American republic from the American socialist state.

No one has described with greater vividness than Rose Wilder Lane what happened in those years, in her book of posthumously published letters, *The Lady and the Tycoon* (1973).[10]

The new generation of Americans has no knowledge of the attempts by courageous individuals in the 1930's to resist the growing tyranny. These battles were lost, and those who fought them have been forgotten. No longer do people resist government loans; most Americans have now traded their liberty for security, or the illusion of security, with all the trappings of Big Government along with it.

It was under Eisenhower that the federal terror on the farms reached its peak. A federal wheat program had already operated for some years, whereby a farmer, if he was to receive federal subsidies for growing wheat, was forbidden to grow more than so much. Many farmers were jailed and fined for growing more, even if they fed it to their own cattle. But Eisenhower initiated in 1954 a *compulsory* wheat acreage allotment.[11]

Over 14,000 farmers were fined during the first year of the Eisenhower wheat program for the crime of growing too much wheat on their own farms. One U.S. district judge was irritated because so many farmers were coming into his court trying to test the constitutionality of the Agricultural Adjustment Act. The judge said he would absolutely not go into the question of constitutionality, and that he was getting so tired of these complaining farmers that he was going to start handing out stiffer sentences to farmers who raised questions of constitutionality.

(2) *OSHA.* If you are a shopkeeper, you are regulated even more. If you have a few employees, you may be able to beat the IRS, and you may be able to beat the inflation, but you are fairly sure to be shafted by the regulatory agencies. Anyone who starts a business, however small, is threatened by a complex of regulatory agencies, any one of which can put him out of business. No court orders or warrants or warnings or notices of intent are required; this is "administrative law." The Occupational Safety and Health Act (OSHA) is empowered "to enter without delay . . . any workplace or environment where work is performed . . . and to inspect and investigate it . . . at any reasonable time" (the OSHA officer decides what is reasonable).

You are the owner of a foundry in Colorado and are fined by OSHA for an excess of silica dust in the air at the plant. You fly in a consultant from Chicago, who establishes that the dust comes from heavy traffic which the city has detoured onto an unpaved road running past the foundry. But you are fined, not only for the dust that is not your fault but for the refusal of employees to wear masks in which they find breathing difficult. You absorb a loss of over $100,000 for that incident.

Or: you satisfy the state inspectors by having ladders with rungs 14 inches apart; the OSHA inspector comes in and says they must be 12 inches apart. The state says guard rails must be 36 inches high, and you comply; OSHA inspectors come in and say they must be 42 inches. Often the two sets of regulations are in conflict, and by satisfying the one you get nailed by the other.

If you try to find out what the regulations are, it will cost you $300 just to buy the documents containing the millions of words in the regulations; no one person could read them all. These regulations, in obscure and vague language, continue to flow in with every passing week; they aren't even indexed, and yet you are expected to have read every word, understand it, and comply instantly.

If you ask OSHA for guidance, you invite trouble. One businessman who asked OSHA for help to determine what he was supposed to do to comply with their regulations was told: "If a compliance officer visits your place of employment, he is obliged to conduct a walk-around inspection. If he finds any alleged violations, it could subject you to assessment of monetary penalties. If you still desire a visit under these conditions, please let us know." (Many similar examples are given in Smoot, op. cit., and Alan Stang, *The Oshacrats.*)

What was the bill of goods which sold the Congress on this particular Gestapo? It was safety; in the name of safety, OSHA was authorized by Congress. Nobody can be against safety, any more than against home, mother, or apple pie. In the name of safety (one aspect, no doubt, of "social justice") a regulatory monster was spawned, with virtually absolute powers and almost total independence of the courts, enforcing edicts conceived in ignorance of actual working conditions, often enforcing practices which are unsafe. Many productive enterprises for which a market demand existed have been hobbled, and others forced into bankruptcy, by the whimsical, capricious actions of this bureaucracy gone wild.

(3) *Energy and the EPA.* Many plant owners report that they have less trouble with OSHA than with the EPA (Environmental Protection Agency).[12]

The impact of the EPA is most strongly felt in the matter of *energy,* and energy is the economic life-blood of America. Without it, industry and virtually the entire economy would come to a standstill. And there is, of course, an energy crisis. But the crisis was not created by the suppliers of energy; it was not created by the Arab boycott; it was created in Washington, by bureaucrats, because the energy-producing industries have been crippled by the regulatory agencies. By the terms of the Environmental Protection Policy Act (1969), anyone, even a Russian agent, can block virtually any new enterprise (on grounds of "interference with the environment") by filing a claim and $100, which will then force the energy producer to write multi-volume memos defending his enterprise and its impact on the environment; and the government bureaucracy, in its turn, may spend years "evaluating" these claims and counter-claims before permitting the enterprise to exist or continue.

There is enough oil off the Continental Shelf east of the U.S. to supply all American oil needs for generations, but any attempt to do so much as explore its possibilities has been blocked by the EPA, as has the development of oil-bearing shale in Colorado and Wyoming. When one tries to fight the regulators, the enterprise of energy production becomes so expensive that it's easier to get the oil overseas. The result is that today Americans pay more for the price of the regulators than for the oil itself—and we are still largely dependent on foreign supplies of oil.

The story of natural gas is less well known, but equally depressing. Prior to 1954, the natural gas industry had already been crippled by regulation in matters concerning the transportation and distribution of natural gas. More than 90 percent of what customers were paying for natural gas was not for the gas itself but for the regulated services of transportation and distribution.

But the regulators wanted more; they also wanted to control the price of the gas at the wellhead. There is, of course, no way to tell in advance the cost of finding and drilling a well; there are almost ten dry holes for every one that turns out to be a producer, and no way to tell how a well will turn out. Nevertheless, after World War II thousands of entrepreneurs were willing to take the risk, and were highly competitive in producing natural gas. Yet in the landmark Phillips Petroleum case of 1954, the Federal Power Commission was asked to set the price of natural gas at the wellhead, because consumers thought that the price they were paying was too high. The case went to the Supreme Court, which in a 5–3 decision decided that the FPC did have regulatory authority over the production of natural gas.

At once, the FPC started giving orders that went far beyond price control. Under FPC rules, a gas producer cannot make a valid contract with customers; any contract can be arbitrarily changed, then or years later, by the FPC. The change may involve reducing the price below the contract price, and the producer must then make rebates. The FPC can order the producer to deliver more, or less, than the contract stipulates; and the producer is not permitted to terminate deliveries until the FPC specifically permits him to do so.

Under such regulations, companies ran away from natural gas production as from a forest fire; it wasn't worth their while to produce natural gas any longer or find new sources of it. Investors looked for more promising places to invest their money. As anyone could have predicted, the production of natural gas declined sharply. Most of the producers simply went out of business; the only ones who survived were the large corporations which produce many things besides gas, and which could turn from domestic to foreign production. The smaller companies were the innocent casualties of the government regulation.[13]

The consumers had been sold a bill of goods in the name of "low prices for necessities" (another aspects of "social justice"). They had what they wanted—cheap natural gas, at FPC-set prices. And at the cheap prices, people started to consume natural gas more wastefully than ever before, as if the supply were infinite. They were like the savages who chopped down the tree in order to get the fruit. As the years went by, shortages of natural gas increased—for which again the producers were blamed. But by this time, the whole thing had become a political football. People wanted natural gas that was both cheap *and* abundant—not realizing that by having the price set so low that nobody could make a reasonable profit discovering new sources, they had cut off the abundance which they also wanted; there was just no way in the world to have both. But the politicians (even those who knew better), ever sensitive to the voters' demands, voted for more and more regulation of the "profiteers"—and the supply went down. So the shortage is now upon us.

Coal and uranium (atomic energy) are the principal keys to America's energy future. The U.S. has almost half the world's coal supply. But both items are so regulated that their full use, in sufficient quantity to take care of our energy needs, is still far in the future. And during that time, we shall be dependent on Arabian and other overseas sources of energy. But if price controls on energy sources are not lifted soon, it would be wise to start putting in windmills, water wheels, and wood-burning heating systems.[14]

What will be the result? More and more energy shortages, avoidable ones, created by Washington. What will be the result of these shortages? Rationing—centralized allocation of energy sources. The bill of goods by which this little package will be sold is: "We have to be sure there's enough for essential industries."

What does centralized allocation of energy sources mean? It means that the U.S. *energy czar* will be a virtual dictator who decides who will get the energy, which region will get the most, which industries will be favored, at what prices, and whose whim will determine what is "essential." This man will be the most powerful man in the nation, and everyone will have to court his favor in order to enable his own industry to survive.

Meanwhile, the energy producers, who *can* produce the needed energy, are already going into other lines of endeavor because survival is too chancy and unpredictable with the increasing regulation. Rather than buck the bureaucratic inertia and unpredictability, they get out of the business they know best.

The nation desperately needs the energy that only they can supply. And yet, the public is quite unaware of all this. People have apparently bought the government line that the energy producers are the villains and that the government is saving them from this villainy—forgetting that the government itself is incapable of producing one iota of energy, but only controls those who do. According to more than one recent poll, the majority of Americans believe that there should be *more* regulation of industry than there is now. Because of this attitude, which also determines what their elected representatives do, there will continue to be a perfectly unnecessary energy crisis which may increase to epidemic proportions. The pattern is: shortages, rationing, riots, famine, economic dictatorship. They will probably not even know that they themselves have initiated the train of actions which led to their destruction. Americans today are like the residents of a town below the dam, who go about their daily rounds not yet knowing that the dam has burst and that their very lives are in jeopardy.

I have only scratched the surface of a discussion of government regulation. There is the FCC, the ICC, and countless other regulatory agencies interfering in the free market, with the result (for example) that plane fare costs twice what it would if unregulated (so the regulated airlines charge double and fly half empty), the railroad industry has already been strangled, and the trucking industry is groaning under the burdens of a host of minute and irrelevant regulations—in many cases doubling the cost of service to the customer. And so on, and so on, *ad infinitum*.

The free-enterprise system, left free of the twin shackles of confiscatory taxation and endless regulation, would yield an ever-expanding technology, with the result that the standard of living would grow by leaps and bounds. It is difficult for us to imagine the enormous prosperity we would experience in a few short years if only the inert hand of government were lifted. If you think of all the small businesses, for example, that are going bankrupt every day in every corner of every city because they can't survive with the taxes and regulations, and how many new inventions can't get off the ground for the same reason (and new medications too, thanks to the FDA), and multiply this by thousands all over the country, you would soon see what a staggering release of human energy there would be everywhere, what gains in employment where workers are now being foregone, what new enterprises would begin and what old ones would expand. The energy of man would be virtually limitless once freedom was restored.

Nor would such an expanding economy entail inflation. With more things produced, and a greater market for them, prices would go down, not up; mass production always means that each item can be produced more cheaply. From 1800 to 1900 in the U.S. there was virtually no rise in dollar prices (except briefly during the Civil War, after which prices came down again). Again, inflation comes from one thing: government's expansion of the money supply.

But because of the actions of government, none of these wonderful things is going to happen. Government will not let go its stranglehold of the economy; and its chances of lowering expenditures (not taxes, but expenditures) is infinitesimally small. The politically popular thing will be done now, and when the deluge comes, whoever is in power will get the blame. What the government takes from you will keep skyrocketing, what is left to you will be less and less, the standard of living will gradually decline, millions of people will be wiped out, millions more will be unemployed with the economy proceeding on half-steam, others will wait at the public till only to discover that it is empty, and most of the victims will have no idea what happened or why—they won't even know that this same tired story has been repeated in history countless times before. And *this* is "social justice"?

V

There is one aspect of the free-enterprise system which the champions of "social justice" constantly attack. "What of the poor?" they say. "Is it just that they should be in their present plight? And what provision is made for them in the free-enterprise system?" I shall make a number of points briefly:

1. Some deserve their fate and some do not; it all depends. There are those who are overtaken with disease or economic catastrophe in spite of every precaution; it is safe to say that these do not deserve their lot. There are others, however, who have thrown precaution to the winds and now use the welfare system to escape the consequences of their own actions. The welfare system does not distinguish between these two categories, and thus encourages the second. Those who have spent their substances in riotous living now tap the welfare system in order to live off those who have been thrifty. Those who can work, but don't and never intend to, are supported by those who do. Who would call this justice? The working people of the nation cannot sustain indefinitely such a tremendous drain on their incomes through confiscatory taxes and inflation. How is the system just to *them?*

2. But let us ignore the waste, graft, inefficiency, corruption; let us ignore even the government's 40 percent handling fee for its various services. What about the *deserving* poor?

People are accustomed to looking through the wrong end of the telescope; instead of asking, "How can I get hold of it, never mind where it comes from?" they could well ask, "How can I ensure that the source of the bounty is sustained and increased?" It is only from the surplus of *production* that handouts can come. Most people simply take production for granted, as if it will continue automatically, no matter what the circumstances; social workers don't mind tapping the spigot as if the supply were limitless—they worry only about the best way to distribute what others have produced. They do not stop to think that if you keep running the water the tank will finally be empty, and that no one will be motivated even to refill the tank.

If the free-enterprise system were to flourish, there would be very little poverty left. Government *creates* most of the poverty that there is in this country. This is too large a statement to go into detail at this late stage (but Clarence Carson's *The War on the Poor* should be compulsory reading). Government soaks up such a large percentage of our income that it keeps everyone fairly poor—poorer, at any rate, than they would be otherwise. It has already created an inflation that will reduce many Americans to bankruptcy. It constantly throws monkey wrenches into the machinery of the free market to divert and stifle human initiative, and regulates to death the businesses that survive. It is a millstone around the neck of every enterprising person who wants to make it on his own. How many thousands of stores and shops and factories would take on more personnel were it not for the taxes and the government-created riskiness of the enterprise, which gives them now the slenderest of margins on which to survive? How many shops that did exist are now closed, thanks to omnipotent government?[15]

What is more charitable, giving a man a handout or putting him back on the production line? Feed a hungry man and he is hungry again tomorrow; put him back to work and he is self-sustaining, with no more need of our charity. Since charity can come only from the surplus of production, charity must be a temporary thing, not a permanent way of living. Our popular morality here is all topsy-turvy. John Paul Getty wrote:

> If I were convinced that by giving away my fortune I could make a real contribution toward solving the problem of world poverty, I would give away 99.5% of all I have immediately. But a hard-eyed appraisal of the situation convinces me that this is not the case. The best form of charity I know is the act of meeting a payroll. If I turned over my entire fortune to a charitable foundation, would it do any more good than I can do with it? The answer is no. However admirable the work of the best charitable foundation, it would accustom people to the passive acceptance of money, and incidentally deprive of their jobs thousands of hard-working people who are associated with me.[16]

Government, by its actions, has put millions of people *out* of the production line by skewering and maiming the enterprises that would keep them in it. Is it not here that the real injustice lies?

3. Now to a moral point. If I see a person who is hungry, I will help him if I can. But does this give me the right to pull a gun on a third party who is looking on, and say to him, "If you don't help this person too, I'll pull the trigger?" But that is just what the government does; government operates not by suggestion or moral suasion but by force and threat of force.

4. But don't people have a *right* to security? Nature offers security to no one. Nature can kill, maim, or impoverish anyone at any time. Nor does the entrepreneur have any security; one or two wrong decisions in the front office and the business goes under. The employees cannot have any more security than the company has; if the company goes bankrupt, the employees cannot continue to be paid. And yet, those who do not work, and depend for their income on those who do, demand a security denied those on whom they depend.[17]

5. But what if all else fails? What of the ill and disabled, who cannot work? I have already described what an outburst of prosperity there would be if government kept its hands off economic enterprises. Most people would have insurance to cover such contingencies, and new competitive companies would arise to meet the demand. And for those who did not, there would be private charities. No, do not laugh; private charities flourish even now, in spite of the confiscatory taxation; they would flourish many times more if that burden were lifted. When you have already borrowed money to pay off the IRS on April 15, you feel less than charitable when the Red Cross comes to your door April 20. You feel, with some justification, that you have already done your bit for charity, inasmuch as the state has largely determined, coercively, to whom you may be charitable and to what degree, even when it goes for causes you find useless or immoral. Without this huge coercive apparatus ripping off our income, the springs of charity would flow again.

VI

It will be fairly obvious by now what I think of socialism; but now, I must make a distinction. I have no objection in principle to *voluntary* socialism, agreed to in advance by all the parties involved. If any group of people wishes to form a socialist enclave, even in the midst of a non-socialist nation, I would say they have every right to do so, and of course, take upon themselves the consequences of their own decision. The standard of living of such a community will probably be much lower, since people are usually more motivated to work and produce when they can thereby benefit themselves, and their motivation decreases when they have to sustain every freeloader in the group. But if that is their decision, I have no right to deny them that preference. I am inclined to think that such arrangements can work only in fairly small groups, whose members are imbued with the same ideals. Be that as it may, I want to distinguish this voluntary socialism from *coercive* socialism, which is imposed by the state on every member of the society, whether he wants it or not. It is this coercive socialism to which I am objecting.

I object to it partly on the grounds of inefficiency and waste, since governmental planning cannot take place without a vast bureaucracy, which eats up a great amount of the national effort in red tape and paper work, and always diminishes production and cruelly distorts distribution.

The U.S. is already more than semi-socialist; and considering the abysmal failure of the bureaucracy in whatever field it has entered, it is simply beyond belief that still further interventions by the state will improve matters. If the state can't even run the post office (and it can't), what could make anyone think that it could run something as complex as agriculture, energy, education, industrial production, the whole intricate machinery of the market? Coercive socialism leads only to underproduction, maldistribution, a lowering of the standard of living, and in the end to splendidly equalized destitution.

But the other reason for opposing coercive socialism is a moral one. Why should I or anyone else have to submit to a system which takes the most fundamental economic decisions (often of life or death) out of my hands and places them in the hands of government officials—to whom, to put it mildly, there is no evidence that such decisions can safely be trusted?

Is planning needed? Yes—by each individual, to plan the course of his or her own life. If it is replied that people are too fallible or too ignorant, etc., then what of the politicians and bureaucrats who rule our lives in a socialist state? Are they not fallible human beings too? What superior wisdom entitles them to manage the lives of others? If history teaches us anything, it is that men placed in positions of power over others will become worse, not better—and that power corrupts, and absolute power corrupts absolutely.

The fact is that any success in planning the lives of other people is highly unlikely and usually impossible. The people being regulated are not even known personally by the regulators; the regulators have no knowledge of the specific and highly individual problems of those they undertake to regulate.[18]

Some time ago, I was engaged in a discussion with a prominent member of the Socialist Labor Party. After he had described the alleged merits of his system, I asked him, "I have described to you what my laissez-faire society would be like. If, in my laissez-faire society, you and others wanted to form a socialist enclave, you would be perfectly welcome to do so, as long as you didn't force anyone to join. Now, in your society, could I 'opt out' in the same way? Suppose I wanted no part of your socialist paradise. Could I say, 'No, thanks, I want no part of your Social Security plans or your national health plans or your central planning bureau. I don't trust them, I want no benefits from them, and I choose not to pay into them, either'? Could I do that?"

He thought for a moment, and then he said, "Well, we all have to pull together."

"And what if I refuse?"

"Then we'll have to *make* you do it."

That answer told me what I wanted to know. And there lies the difference between any socialist system and the free-enterprise system. In the socialist society, the leaders could not permit widespread dissent; and in the end, they would have to shoot me. But in the free-enterprise society, they would go free—free to do whatever they wished, as long as they did not coerce anyone else.[19]

No, private enterprise may not create a system of ideal justice; but it comes infinitely closer to it than does the rule of government. No economic system can guarantee justice in every individual case; but omnipotent government *can* guarantee *in*justice. Government promises something for nothing, and delivers in the end nothing for something. It places our income, our privacy, our individuality, our very lives at the mercy of its various agents, to dispose of as they decide. Government, which is supposed to protect us, has looted, robbed, and killed a thousand times more than all the criminals in history put together. The free market, whatever its limitations in achieving justice in individual cases, is a paragon of justice compared with this monster leviathan. If the state has taught us anything at all about justice, it is by presenting us with its very antithesis.

NOTES

1. There was a third event also: lest we not heed the lessons of the first two, Gibbon's *Decline and Fall of the Roman Empire*, published in 1776, told us clearly enough what would happen.
2. Cf. Rose Wilder Lane, *The Discovery of Freedom* (New York: The John Day Co., 1943), pp. 237–39.
3. Milton Friedman, in *Harpers,* February, 1973.
4. Leonard Read, *Clichés of Socialism* (Irvington-on-Hudson: Foundation for Economic Education, 1970), pp. 165–66.
5. Cf. F. A. Hayek, *The Constitution of Liberty* (Chicago: University of Chicago Press, 1960), p. 97.
6. *Idem.*
7. Ibid., p. 98.
8. Gary Allen, *American Opinion,* July, 1976, p. 109. See also Irwin Schiff, *The Biggest Con* (New Rochelle, N.Y.: Arlington House, 1976).
9. *Newsweek,* June 9, 1975, p. 26. See also C. V. Myers, *The Coming Deflation* (New Rochelle, N.Y.: Arlington House, 1976).
10. See especially pp. 168–70. Also J. Hospers, *Libertarianism* (Los Angeles: Nash Pub., 1971), chapters 3 and 4.
11. Cf. Dan Smoot, *The Business End of Government* (Boston: Western Islands, 1973), p. 83.
12. Genuine environmental problems should be taken care of not by regulatory agencies but by taking the polluters to court, treating such matters like other cases in tort. (See J. Hospers, *Libertarianism,* chapter 9.)
13. See the series of articles by Eugene Guccione in *Mining Engineering,* September 1974 *et. seq.* and *Reason,* October 1977; also, Murray Rothbard, "Carter's Energy Fascism," *Libertarian Review,* July, 1977.
14. See the Gary North Newsletter, *Remnant Review,* August 6, 1975.
15. Cf. Henry Hazlitt's excellent book, *The Conquest of Poverty* (New Rochelle, N.Y.: Arlington House, 1973); also his *The ABC of Inflation* (Lansing, Mich.: Constitutional Alliance, Inc., 1969).
16. John Paul Getty, *Saturday Evening Post,* May 22, 1965, p. 10.
17. Cf. Rose Wilder Lane, *The Discovery of Freedom,* p. 60.
18. See Herbert Spencer, *The Man vs. the State* (London: Williams and Norgate, 1884), pp. 41–43, 122. Also J. Hospers, *Libertarianism,* chapter 7.
19. Cf. Roger Macbride (ed.), *The Lady and the Tycoon* (Caldwell, Idaho: Caxton Printers, 1973), pp. 330, 332–33.

III

ETHICAL CONSIDERATIONS IN CIVIL REMEDIES

6

IS PREFERENTIAL TREATMENT FOR RACIAL MINORITIES AND WOMEN JUST OR UNJUST?

WILLIAM T. BLACKSTONE

Let me begin by stating the general problem with which I am concerned in this essay and by drawing certain distinctions which will permit me to narrow the focus of my concern in a reasonably clear way.

My concern is with whether or not the preferential treatment of minorities and women—often called reverse discrimination—is just or unjust. For a number of years—especially since the 1964 Civil Rights Act, later amended by the Equal Employment Act of 1972 and the Affirmative Action Program which came into existence basically with Executive Order 11246—there has been a good deal of preferential treatment of women and racial minorities. Sometimes the preference has taken a weak form, in which the minority member or woman is given preference in contexts where that candidate is equally qualified as a majority candidate. On other occasions, preference is given even when the minority member or woman is less qualified. There is little doubt that this has occurred in practice, especially where numerical goals and quotas have been imposed on institutions and businesses. And the not unexpected response has been that this preferential treatment for some is discrimination against others. Majority members have claimed that they are being discriminated against because of their race or sex and that no matter how well intended such practices and policies may be, their net effect is invidious discrimination against majority members of society. Thus, we now have numerous charges of reverse discrimination lodged by majority members. The DeFunis case, declared moot by the Court, is probably one of the most famous.[1] More recently, AT&T, after having signed a consent degree with the federal government to hire more women, was successfully sued by a man who charged that he was unfairly denied promotion to a supervisor's job because of discrimination in favor of women.[2] Numerous claims of reverse discrimination have also occurred where seniority rights have come into conflict with the objectives of equal employment opportunities. Possessors of seniority rights under labor union agreements argue that those rights should not be overridden by policies of racial and/or sexual preference—that last hired, first fired is proper regardless of the differential impact on minorities and women, who have been recently employed in such unionized firms.

In this essay, I want to discuss whether preferential treatment for classes of persons like blacks and women is just or unjust. Within the confines of our commitment to justice, what can be done, what ought to be done, in situations where not only individuals but entire classes of persons continue to suffer the effects of past institutionalized injustice and where racist and sexist practices continue to prevail in fact, even after declared illegal? Within our commitment to justice, what can be done to assure genuine equality? The social facts which are obstacles to that equality are not only the persistence of practices which once were legal but now are not, but also the fact that many members of these classes of persons who were discriminated against in the past and denied certain opportunities now are not in a position to compete on the same meritocratic criteria with non-disadvantaged majority members. And further, the children of these persons often have inherited major disadvantages as a result of the disadvantages of their parents.

What can be done, consistent with justice for all, to rectify this? Surely benign neglect of the effects of the sins of the past will simply help to perpetuate inequality. To say to someone who has been disadvantaged by past injustices of the state, "Get in there and sink or swim with the rest of us. The door is now open, compete with the non-disadvantaged majority"—this may be unfair, given the accretions of history. That is, mere non-discrimination at this stage may vitiate the attainment of equality, even though many women and blacks would be willing to settle for that. Thus, it is argued by many that more positive, affirmative action is required to bring disadvantaged minorities and women into the mainstream, where they have equal status with the non-disadvantaged middle-class white male and where they can compete within a meritocratic system on equal terms. Non-discrimination, even if instantiated in practice—and many institutions and businesses persist in covert racist and sexist practices in employment—would not be enough to assure justice and equality. It is argued by many that the affirmative action which goes beyond mere non-discrimination must include preferential treatment for those classes of persons who suffered institutionalized injustice in the past, and that we are now required by justice to discriminate in their favor, using the class-identifying characteristics of race and sex. It is not a mere pejorative use of a term to call this "reverse discrimination."

To make sure that my restricted topic is seen as I have circumscribed it, let me set off several other questions with which I am not concerned here. (1) Do the recently enacted civil rights laws require or permit preferential treatment on the basis of sex or race? Some argue that such preference is absolutely forbidden by those laws and that the implementers of the laws have initiated policies of reverse discrimination by requiring quotas. I will not discuss this question. (2) Nor will I discuss the question of whether reverse discrimination is morally justified, except to the extent that justice is a part of morality. Morality is a much broader normative framework for assessing reverse discrimination. It surely includes justice, but, in my opinion, morality as a whole involves concerns and principles which are not reducible to justice and equality. Nor is justice reducible to these other moral concerns, like social utility. We cannot settle the justice of an act or policy by finding out whether it maximizes social utility. These two questions are important ones, but they are not our concern here.[3]

Now, in asking whether justice requires or permits reverse discrimination, we must have some reasonably clear sense of what we mean by justice, and to get this reasonably clear sense, we must make several distinctions. One distinction is that between what we might call legal justice and moral justice. Another is that between compensatory and distributive justice.

By legal justice, I mean that which is required by the law or the Constitution, and, of course, we must distinguish between the law and the Constitution. Laws may violate the Constitution, and the legal justice with which I am concerned is not conformity to law but conformity to the Constitution. Moral justice may or may not be embedded in either. To the extent that it is, there is an overlap between legal and moral justice. There is often that kind of overlap, so that there are close parallels between arguments concerning the moral justice of an act or policy and the legal justice (constitutionality). But again, it is possible that constitutional justice may fly in the face of moral justice, in the same way that certain laws passed by legislators may fly in the face of constitutional requirements. My concern in this essay is with legal justice in the sense of constitutionality and, under this rubric, simply with what is permitted or required by the Equal Protection Clause of the Fourteenth Amendment. There are different interpretations of what is required or permitted by equal protection, and that will be our primary focus in discussing the justice of reverse discrimination.

Like legal justice in the sense of constitutionality, what constitutes moral justice depends on the moral framework adopted and on the interpretation of that framework. And just as there may be different constitutions and different interpretations of what is required by provisions of those constitutions, so there may be different moral frameworks of justice and different interpretations of what is required by those frameworks. Aristotle, Marx, Mill, and Kant all offer fundamentally different moral frameworks of justice. Whether there is such a thing as *the* moral point of view, which embodies in some sense neutral methods and procedures for devising and deducing *the* moral framework of justice is questionable.[4] The most recent systematic attempt to derive a substantive theory of moral justice from a presumed neutral and mutually acceptable procedure has been declared by many to be circular and question-begging. I am referring to Rawls's mammoth and monumental *A Theory of Justice*.[5]

In any case, we cannot here take on these broader questions. I will simply appeal to a sense of justice as equality, in which equality means equal status as persons or equality of consideration. Like the Equal Protection Clause, moral equality or equality of consideration is subject to interpretation. Both are open-textured concepts, as Paul Freund indicates when he speaks of the Equal Protection Clause as a moral standard put in the wrappings of a legal command. There is no doubt that that clause has permitted the Court to mold the nation's thinking on social justice.

With this affinity between the Equal Protection Clause and the moral principle of equality before us, I now want to concentrate basically on interpretations of the Clause, and indeed, as Freund indicates, this may also amount to interpretations of moral equality. We want to see the implications of different interpretations for the question of the justice of reverse discrimination as a policy. Neither the doctrine of equal protection nor the moral principle of equality can be said to be longstanding features of our Constitution or of our Western moral tradition. The former became part of the Constitution only with the Fourteenth Amendment; the latter, though present in Stoic and Christian thought in some form, has hardly been a major part of the Western moral landscape. Both are late emergents, and both are still evolving in meaning—as Freund indicates.[6] The question before us is whether some recent evolution of these doctrines should become more or less permanent features of our moral and constitutional landscape.

Now earlier, I mentioned the distinction between compensatory and distributive justice. Both of these types of moral justice are subsumable under equality or equal protection. So they are not interpretations of equal protection as a principle but different kinds of equal protection. Their impact on the question of the justice of reverse discrimination, then, has more to do with the logic of the concepts of compensation and distribution than with the more generic notion of equality or equal protection. We must, however, distinguish these types of justice and their use in equal protection arguments before moving on to generically different interpretations of equal protection, because both are invoked as rationales for preferential treatment of women and racial minorities.

Aristotle introduced the distinction between compensatory or rectificatory justice and distributive justice in Book V of the *Nichomachean Ethics*. The former involves a reparatory or corrective transaction between one person or party and another. The objective is that of restoring the equality or state of affairs which existed prior to the injury of one party by the other. Compensation is to be exacted from the perpetrator of injury and awarded to the injured party, and the benefit bestowed on the latter is to be proportional to the difference created by the act of the former. Distributive justice, on the other hand, involves the proper criteria for the distribution of goods, services, and honors in society. In both cases, Aristotle thought basically of justice as involving proportional treatment of individuals, depending in the case of compensation on the scope of the wrongful injury and, in the case of distribution, on the possession of (what he considered to be) relevant characteristics (aristocratic status, intellectual and moral virtues, and the like).

The bases for current demands for preferential treatment of women and racial minorities are both compensatory and distributive justice. Preferential treatment is now required, it is argued, in order to compensate these classes of persons for injuries due to past institutionalized injustices. And preference is required in order to assure distributive justice now—given the cultural lag and recalcitrance to equality of consideration for all—and for future generations. When the compensation for past injuries is paid to these classes and when the recalcitrance to equality of consideration is overcome—that is, sex and race no longer are invoked as relevant where they are irrelevant—then it may be possible to return to—or should we say "adopt," since we were never there—a principle of non-discrimination.

Unlike the position of Aristotle, these rationales demand compensation and distribution on a class or group basis, not merely on an individual basis. Classes of persons have been harmed, it is argued, and classes of persons must be compensated; and the whole distributional scheme must take not just the fact of injury but current recalcitrance to equality into consideration in a policy to assure equality.[7] To this end, it is argued, the appeal to class characteristics of race and sex—are justified. Furthermore, these compensatory and distributional moves are requirements or obligations of society as a whole, not simply debts owed by individuals who wrongfully injured others.

What, then, are we to say of all this? Is preferential treatment for women and racial minorities just or unjust? Obviously one's conclusion depends on the theory of justice—in this case, the interpretation of equal protection, which one adopts. Ronald Dworkin has recently warned us that "we must take care not to use the equal protection clause to cheat ourselves of equality."[8] Obviously he has a certain interpretation of the generic meaning of that clause in mind. In fact, he distinguishes between the right to be treated as an equal and the right to equal treatment. The right to be treated as an equal Dworkin sees as the fundamental meaning of equal protection. It requires that each person be treated with the "same respect and concern." The right to equal treatment—which I will call the right to non-discrimination—is the right to the equal distribution of some resource, opportunity, or burden. This right, Dworkin insists, is "derivative." It is entailed by the right to be treated as an equal in some circumstances but not in all. In the case of Marco DeFunis's exclusion from the University of Washington law school, for example, the fundamental right at stake for DeFunis and everyone else was the right to be treated as an equal, not the right to equal treatment. DeFunis's application, his interests, his potential losses and gains were not ignored by the admissions committee. They were treated "as fully and sympathetically as the interests of any other," which is what equal protection requires, Dworkin holds. But equal protection does not require equal treatment—that is, equality of distribution. In fact, it may require, in some circumstances, discriminations which deny equal treatment. The discrimination which took place in the DeFunis case, Dworkin holds, was not arbitrary, for it did not violate DeFunis's right to be treated as an equal, and the racial classification and discrimination on the basis of race was, under present circumstances, required to fulfill the right to be treated as an equal of members of minority groups.

I will return to this distinction and Dworkin's argument in a moment, and try to assess its weight in justifying preferential treatment for blacks and women. But one further distinction and premise invoked in favor of preferential policies I want to lay out for inspection. It is a distinction and a premise which Professor Cornelius Golightly thinks underlies affirmative action requirements.[9] The distinction is that between ascribing rights to groups or classes and ascribing rights to individuals. The premise is that we should accord rights to entire classes—blacks, women, American Indians, and so on. Not all classes—only those that have been unable to make it into the American mainstream because of past and current discrimination. The Fourteenth Amendment vests rights in individuals. Its thrust is non-discrimination. The affirmative action requirements, on the other hand, with their emphasis on the "underutilization" of certain groups and the use of goals and timetables—which in practice sometimes become specific quotas—vest rights in groups. These are two different "political moralities," Golightly states, and they simply clash in certain circumstances; for the emphasis on group rights—a kind of "tribal morality"—implies that individual rights can be overridden in the interest of group goals. Non-members of favored groups can be discriminated against simply because of that non-membership (because of their sex or race.)

The group or tribal morality which the U.S. government has imposed within affirmative action guidelines—and it exists right alongside an individual-rights emphasis in the same guidelines—is seen as a move forced upon us by the recalcitrance of our society to bring previously legally oppressed groups within the mainstream of equal opportunity on an individual basis. It is temporary. Affirmative action programs are to be phased out as the accretions of racism and sexism are overcome, and the full individual-rights emphasis will return.

Now, applying this line of reasoning to the DeFunis case, what do we get as an analysis? First, we must note that *DeFunis* did not derive specifically from affirmative action guidelines, but the logic of the issues is the same. Let us accept the view that there was no discrimination against DeFunis in relation to the majority applicants but that there was in relation to the limited seats set aside for minority members. And let us suppose that the rationale of the Admissions Committee included, along with other things, the objective of overcoming both the effects of past institutionalized racism and the persistence of racism in our society. Let us suppose that the major thrust of the "compelling state interest" argument in the *DeFunis* case (Washington Supreme Court) is this objective, that in effect the Admissions Committee attempted to balance this objective with the principle of non-discrimination applied in the case of the rest of the seats available in the first-year class.

I say, "Let us suppose all this," because I do not deny that another rationale is possible or that several rationales were operative. Certainly the possible prejudicial nature of the standard indices of ability—the LSAT and PGPA—may have played a key role. If this were the major rationale for the differential treatment of minority members, and if it could be shown that the same standards of merit or ability were applied to all applicants, majority or minority, the only difference being that the indices of ability differed, then one could justifiably infer that the principle of non-discrimination guided the actions of the Committee in all cases. This rationale for differential treatment is exactly what Justice Douglas leaves open as a reading of the *DeFunis* events, and it is why he declares in his dissent that he cannot say, on the basis of the facts at his disposal, that there was racial discrimination against DeFunis.[10] The use of racial classification, as found in *DeFunis,* is perfectly consistent with the principle of non-discrimination and with the Constitution, as long as race itself is not the dispositive factor in allocating admission but simply a tool to enable the Admissions Committee to discover minority members who meet the required criteria of ability and potential, but as measured by non-standard indices; that is, as long as it is a tool for assuring equality of treatment regardless of race.

What is problematic about the *DeFunis* decision in the Washington Supreme Court, in the sense of conformity to the requirements of either constitutional justice or moral justice, is the reading of the case in which there is an overriding of the principle of non-discrimination or equality of treatment for a majority race applicant in order to effect greater equality for minority members. If the appeal to a group characteristic, race, was the dispositive factor in the exclusion of DeFunis, then perhaps his constitutional and moral rights were violated. I say "perhaps" because if Dworkin is correct, what was denied him was the right to equal treatment, not the right to be treated as an equal. If this is true, and if the latter is the generic meaning of equal protection and the former only derivative, as Dworkin argues, then DeFunis's right to equal protection as guaranteed by the Fourteenth Amendment was not violated. I have promised that we will return to this distinction and this argument. And we will. But first, let me make several observations on matters related to differential treatment on the basis of class or group characteristics.

The first point is this, and it is emphasized strongly in Boris Bittker's *The Case For Black Reparations:*[11] Who comprises the class of blacks? The answer to the analogous question about women is somewhat simpler, in spite of sex-change operations—though I too would be wary of meeting Dr. Renée Richards on the courts if I were a woman. But, in the case of blacks, interracial breeding has reached such proportions that a huge number of our population could claim to be black; and, as Bittker points out, if the benefits were great enough from class inclusion, many of us would rise to the occasion. There are not inconsiderable difficulties in circumscribing the boundaries of the class. Secondly, though compensation and distribution of resources on a group basis are by no means conceptually absurd and have occurred historically, the normal presupposition of such occurrences is that there is a close empirical connection between benefit or injury of individuals and membership in the group. As Bittker points out, Indian tribes meet this criterion. But he suggests that blacks do not. Even less do women. There are no singular identifiable spokesmen for blacks as a group, nor for women.

Bittker, then, suggests that there are serious problems of ill fit in treating blacks as a class. Others have argued that there also are serious problems of ill fit in compensatory and distributive policies which operate on the basis of race or sex. They often result in gross injustice. This is so for several reasons: The logic of compensation requires that compensatory redress be proportionate to the injury suffered and that the redress be paid by the party guilty of the injury. Blanket policies of preference on the basis of sex or race have the effect of exacting redress and awarding compensation arbitrarily. Redress is often exacted from majority members who themselves are not guilty of any wrongdoing, and compensation is often awarded to minority members and women who have escaped the effects of past institutionalized injustice and who in fact have advantaged backgrounds when contrasted with many majority whites.

Those who argue for preferential treatment on the basis of sex or race agree that these injustices occur, but they say that it is a price which must be paid to effect overall social justice. Some injustice is justified on the grounds of the overall just result. Ideally, a policy directed only at individuals would be the best instrument. But, they declare, such a policy would be administratively impossible. Research on individuals, the nature and extent of their injury, the guilty party, the amount of compensation, and the like would be impractical and would in effect halt compensatory action. For most of the disadvantaging injuries cannot be directly attributed to culpable individuals in the here and now, and yet, the disadvantaged condition of minorities and of many women is plain to see. Statistical evidence of those persons who are in and out of the mainstream of American life confirms it.

Is administrative convenience an acceptable reason for overriding the rights of majority members? For arbitrarily exacting compensation and awarding benefits? And what can be a principled response under such a system when the rights of members of one group—say, blacks—clash with the rights of another preferred group—say, women or Chicanos?

Is there to be a hierarchy of group rights? I mention these issues here simply to lay them aside. They are formidable problems for those who would press for group morality or for social justice on a group basis, but they are not my central concern here. That concern is with whether the violation of the principle of non-discrimination in any DeFunis-like context can be said to be not unjust and consistent with our constitutional and moral commitment to equal rights or equal protection. That is, is racial (or sexual) discrimination both constitutional and just in some circumstances?

Obviously everything hangs on one's view of what is required by the Constitution—in this case, the Equal Protection Clause—and moral justice. The thrust of the decision of the Washington Supreme Court in the case of *DeFunis* is that "compelling state interests" justify not only racial classification of students but racial discrimination in favor of minority students, at least for limited seats in the law class. The brunt of the "compelling state interest" might be summarized as the moving of minorities into the profession of law in order to bring about long-term racial justice.

Actually, we should distinguish two interpretations of the "compelling state interest" argument for classifications and policies which favor racial minorities and women. One interpretation appeals to equal justice for all or long-term, overall social justice as grounds. The reading of the "compelling state interest" in DeFunis, given above, fits this form. So also does the rationale offered by Justice Brennan for a statute which included a classification favoring women. He stated: "I agree that in providing a special benefit for a needy segment of society long the victim of purposeful discrimination and neglect, the statute serves the compelling state interest of achieving equality for such groups."[12]

A second reading of the "compelling state interest" argument is that preferential policies are justified in terms of the overall benefits for society as a whole. Such policies improve the economic and social condition of the preferred groups. They end civil disorder. They maximize social welfare and social happiness. This is a utilitarian reading of the "compelling state interest" argument.

Now, many deny that these results will follow from preferential policies. For various reasons, they believe the net effect of such policies will be negative. Justice Douglas, for example, suggests that preferential policies will place a stigma of inferiority on minorities and, to that extent, are self-defeating.[13] I do not want to engage these empirical claims here. But I do want to note that a utilitarian argument of this type, if accepted, opens the door for the overriding of any number of individual rights which we think of as protected by the Constitution and equal protection. And the utilitarian knife can cut both ways—against the rights of minority and majority members. It may be this utilitarian reading of the "compelling state interest" argument which Justice Douglas had in mind when he said in his DeFunis dissent: "If discrimination based on race is constitutionally permissible when those who hold the reins can come up with 'compelling' reasons to justify it, then constitutional guarantees acquire an accordion-like quality."[14] Justice Douglas points to the fact, admitted by the Washington Supreme Court itself (which ruled against DeFunis), that a policy of preferential admissions is "not benign with respect to non-minority students who are displaced by it." On the other hand, Justice Douglas's objection may also be to the argument which pits DeFunis's right to non-discriminatory treatment against the right of oppressed minorities to have the burdens of the social sins of the past removed. That is, his worry about the "compelling state interest" argument may include the reservation that such arguments should not override the rights of a DeFunis in the process of rectifying the sins of the past and assuring equality for minority members. It is not merely that two wrongs do not make a right. It is that the result of such moves is that of undermining the entire principle of equal protection.

Now, obviously Dworkin does not accept this conclusion. He does flatly reject utilitarian arguments which attempt to justify the discriminatory disadvantaging of members of one race on the grounds that such discrimination maximizes social utility. Even if such a policy did maximize utility, it is unacceptable because it is unfair. Other utilitarian arguments may not involve unfairness and may justify preferential treatment of minorities. But the argument on which Dworkin rules is a non-utilitarian one. The right to equal treatment, he declares, may be overriden by the right to be treated as an equal. So, in the case of DeFunis, his right to equal treatment or his right to non-discrimination, especially in contexts where his "vital" interests are not at stake, may be overridden by the right to be treated as an equal of others.[15] Equal protection requires, as Dworkin puts it, that everyone be treated as an equal, and unless perferential treatment is given to racial minorities and women (which may involve discrimination against non-preferred persons and overriding their right to equal treatment), large numbers of persons will not be treated as equals. What we are being asked to do by Dworkin is to pit a higher-order right or principle, which he takes to be the generic meaning of equal protection, against a lower right which is derivative of it. We are asked to accept the conclusion that even if DeFunis's right to equal treatment were violated, he was not treated unjustly or unconstitutionally because that violation was premised on the basis of the right of everyone, including DeFunis, to be treated as an equal.

Does this argument hold water? I do not think it does, because it seems to me that the distinction between the right to equal treatment and the right to be treated as an equal breaks down. Are we treating DeFunis as an equal when we discriminate against him and exclude him on racial criteria (assuming, for the moment, this reading of the facts of *DeFunis*)? Are not the interests of anyone who is excluded from admission, employment, or promotion simply because of his race being relegated to secondary status? It seems to me that they are. Such a person is not being treated as an equal. His interests are being adversely affected simply because of his race. One must surely agree that the exclusion of DeFunis on racial grounds would not have injured him to the extent that past racial discrimination injured blacks. Both the stigma and the effects of the disadvantaging would have been much less. (DeFunis was admitted on appeal and had practically finished law school before the Court heard his case.) But the principle behind his exclusion, despite the good intentions and the assumed temporary nature of the policy, was the same for DeFunis as it had been for blacks earlier, e.g., invidious discrimination.

It must be granted that the uses of racial classification are not always invidious. Race can surely be considered in order to effect social justice, as occurred in the school desegregation cases, *Green* and *Swann*. But in these cases, race was not used to apportion some benefit to members of one race and exclude those of another but to assure the possibility of equal benefits for all. In the case of DeFunis, race was used to exclude him from certain benefits in order for the Law School to pursue other compelling objectives.

I am suggesting two things. Other perhaps than national survival itself (and even here, the greatest caution must be observed in invoking this ground), there is no compelling state interest which justifies racial discrimination. To embrace such discrimination is to give up the guts of equal protection. Which brings us back to Dworkin's distinction. What I am saying is that the purported distinction between the right to equal treatment and the right to be treated as an equal simply collapses. If what Dworkin means by the right to equal treatment is the right to the *same* treatment—that is, identical treatment—then we would surely grant him the distinction between the right to equal treatment and the right to be treated as an equal; and we would grant that the right to the *same* treatment is derivative. Sameness of treatment is required only when it is entailed by equality of treatment.[16] This is in fact the commonplace interpretation of the principle of non-discrimination. Non-discrimination does not require that we ignore relevant differences which justify the differential treatment of persons. It requires, on the contrary, that we attend to them. But this commonplace is not what Dworkin is pressing. He seems to be arguing that the right to non-discrimination can be overridden by the right to be treated as an equal. What I am suggesting is that there is no difference between these rights. Overriding DeFunis's right to non-discrimination is equivalent to overriding his right to be treated as an equal, a right which is now fundamental to our Constitution and our way of life.

The rejection of Dworkin's argument does not preclude one's giving priority in certain contexts to one right as opposed to another, both of which are grounded in moral justice or in the Fourteenth Amendment. The right to non-discrimination, for example, as grounded in equal protection might in some contexts be deemed less important than some other right—say, a welfare right. Put in somewhat different language, the right of one person not to be economically disadvantaged might justify the violation of the right to the uniform application of meritocratic standards of access to positions of another. This is in fact argued by Thomas Nagel.[17] Under ideal conditions, these rights would not conflict. But if they do, it is not irrational, and it may be justified, to adopt the preference just stated. Such preference would be based on the judgment—and perhaps a theory—that economic justice has priority over non-discrimination *in the sense* of the uniform application of standards of access to positions. But this priority, if applied independently of one's sex or race (even if sex or race is used to identify those who are economically disadvantaged), does not violate the general principle of equal protection or non-discrimination, because a relevant feature (economic disadvantage) for differential treatment is present. The problem is justifying this priority on constitutional grounds. I intend here simply to leave open this priority. I have no philosophical or jurisprudential theory ready at hand to defend it. The position of Rawls in his *A Theory of Justice*, under his "special conception" of justice, would yield the opposite conclusion, for non-discrimination would fall under either the principle of equal liberty or that of equality of opportunity (or both), and both have lexical priority over the difference principle or welfare rights. Neither the Rawlsian priority nor that of Nagel invoke sex or race as the dispositive factor in granting the priority.

It is plain, from what I have said above, that I reject two premises which function as premises in arguments for the preferential treatment of persons on the basis of sex or race (I am not here assuming that such treatment reflects the laws undergirding Equal Opportunity and Affirmative Action). One is that group rights exist and that they override individual rights. Another is that the moral and constitutional principle of equal rights or equal protection permits or requires the denial of any individual's right to non-discrimination. What, then, do we do about large numbers of individuals who belong to those loosely defined classes of persons who are disadvantaged because of past injustice, who are still discriminated against on the basis of their sex and race, illegal though it may be?

The answer is that we vigorously, and with adequate national resources, attend to the conditions which have disadvantaged, and which continue to disadvantage, any person of any race or sex and overcome the obstacles which prevent equality of opportunity. In some cases, this effort requires the application of the principle of compensatory justice, where specific individuals or parties have injured others and where justice requires that the guilty party rectify the injury. This holds in cases in which racial or sexual discrimination against minorities or women can be traced to identifiable individuals or corporations. In other cases, perhaps the greatest number, the quest is simply for distributive justice for disadvantaged persons who are disadvantaged because of past officially permitted practices, but where there is no identifiable individual or party guilty of explicit discrimination against them. The disadvantaged condition, though not explicitly traceable to a guilty discriminator (unless we treat *inaction* on the part of individuals or the state as the cause of the disadvantaged condition), still effectively prevents many from enjoying genuine equality of opportunity. We know that many blacks and members of other racial minorities suffer from disadvantages caused by state-supported discrimination. And women too, though that is another kettle of fish in terms of causal factors. Now, knowing that history and knowing the current conditions, we are perfectly justified in invoking racial and sexual categories in looking for disadvantaged persons. Given the facts of history, we are entitled to assume that many will be found there. But we are not justified in using those categories to exclude majority members who, for other reasons, may also be disadvantaged and who also deserve differential treatment to provide them with equality of opportunity. The relevent fact justifying differential treatment is the fact of injury or the disadvantaged condition, not one's sex or race. The latter are simply handy tools for identifying persons who are disadvantaged. Nor can those categories be used to disadvantage majority members who have not historically suffered from racial or sexual discrimination.

This amounts, in my opinion, to the enforcement of non-discrimination. That is what is called for by the 1964 Civil Rights Act, the Equal Employment Act of 1972, and the affirmative action programs grounded in Executive Order 11246. Affirmative action simply requires that we attend to the relevant differences between minorities, women, and majority members of society and focus on justified differential treatment in order to provide equal opportunity for all, not that we discriminate in favor of anyone.

In conclusion, I want to press a somewhat different line concerning the problem of the perpetuation of past unfairnesses into the future. We must stop that perpetuation, and I agree fully with the objective of affirmative action programs, if not with the use of quotas and numerical goals as a means. We have done far too little as both individuals and as a society to overcome the current effects of past discrimination and the persistence of racism and sexism in our society. I have suggested that a strict enforcement of the principle of non-discrimination, which includes attention to relevant differences between persons such as disadvantaged conditions, will go a long way toward solving the problem. But it seems to me that we ought to emphasize not just that portion of the Fourteenth Amendment which calls for non-discrimination but that part which calls for "no deprivation of life without due process of law." Government has been far too inactive in formulating programs which will assist in overcoming disadvantaged conditions which are not due simply to discrimination, past and present.

Why should the bulk of efforts at remedy fly under the flag of discrimination? When government has the resources to overcome disadvantaging conditions not directly attributable to discrimination, is not benign neutrality itself an injustice? An affirmative answer to this question is implied by the Higher Education Guidelines, which properly ask that we make "additional efforts to recruit, employ and promote qualified members of groups formerly excluded, *even if that exclusion cannot be traced to particular discriminatory action* on the part of the employers" (my italics). Neither I nor those Guidelines would have us ignore the large amount of ongoing racism and sexism in our society, but we don't have to identify discriminators in order to have a rationale for policies to overcome disadvantaged conditions. The alleviation of those conditions is supported by both an equal protection and a due process basis. The objective of the Fourteenth Amendment is not simply that of preventing arbitrary state discrimination; it is that of acting positively to prevent deprivation of life. Disadvantaged conditions often are rooted in sheer deprivation, not discrimination, or at least, not discrimination in which culpable discriminees can be charged for remedy and compensation. And yet, there is surely collective responsibility for remedying those conditions.

Frank Michelman has suggested that we look to the Due Process clause as a constitutional basis for rights for the poor—for welfare rights.[18] I am not by any means suggesting that we reduce discrimination to poverty, but I am suggesting that "no deprivation of life without due process of law" might serve as a base for the rights of the disadvantaged of any race or sex, whether the disadvantage is rooted in poverty or other causes. Affirmative action programs rooted in both equal protection and "minimum protection"[19] should be far more effective programs in moving the disadvantaged into the mainstream of American life, far more effective and far more just than programs involving reverse discrimination.

NOTES

1. *DeFunis* v. *Odegaard;* 94 S. Ct. 1704 (1974).
2. See the report in *Newsweek,* June 21, 1976.
3. These questions are treated in my "Reverse Discrimination and Compensatory Justice," *Social Theory and Practice,* Vol. 3, No. 3 (Spring, 1975) and in my "Compensatory Justice and Affirmative Action, *Proceedings of the American Catholic Philosophical Association,* Vol. 49 (1975), as is the issue of the justice of preferential treatment. Here, I extend the analysis of the question of the justice of such policies further.
4. For discussion of the moral point of view, see Kurt Baier, *The Moral Point of View* (Cornell University Press, 1958); Paul Taylor, *Normative Discourse* (Greenwood Press, 1961); William Frankena, "The Concept of Morality," *The Journal of Philosophy,* Vol. 63 (1966); and Hector-Neri Castaneda and George Nakhnikian, eds., *Morality and the Language of Conduct* (Wayne State University Press, 1963).
5. John Rawls, *A Theory of Justice* (Cambridge, Mass.: Harvard University Press, 1971); see Norman Daniels, ed., *Reading Rawls* (New York: Basic Books, 1975), for critical treatments of Rawls's theory.
6. Paul Freund, *On Law and Justice* (Cambridge, Mass.: Harvard University Press, 1968), p. 38.
7. See Paul Taylor, "Reverse Discrimination and Compensatory Justice," *Analysis,* Vol. 33 (June, 1973), pp. 177–182.
8. Ronald Dworkin, "The DeFunis Case: The Right to Go to Law School," *The New York Review,* February 4, 1976, p. 33.
9. Cornelius Golightly, "Justice and 'Discrimination for' in Higher Education," *Philosophical Exchange,* Vol. I (Summer, 1974), pp. 5–14.
10. Justice Douglas's dissent in *DeFunis* v. *Odegaard,* 94 S. Ct. 1704 (1974).
11. Boris Bittker, *The Case for Black Reparations* (New York: Vintage Books, 1973).
12. *Kahn* v. *Shevin,* 416 U.S. 351, 358–9 (1974).

13. Justice Douglas, op. cit.
14. Justice Douglas, op. cit.
15. Dworkin states: "Individuals may have a right to equal treatment in elementary education, because someone who is denied elementary education is unlikely to lead a useful life. But legal education is not so vital that everyone have an equal right to be admitted." Op. cit., 30. There is no discussion of criteria for "vital" interests.
16. For discussion, see my "On the Meaning and Justification of the Equality Principle," *Ethics,* Vol. 77, No. 4 (July, 1967), pp. 239–253.
17. Thomas Nagel, "Equal Treatment and Compensatory Discrimination," *Philosophy and Public Affairs,* Vol. 2 (Summer, 1973), pp. 348–363.
18. Frank Michelman, "The Supreme Court, 1968 Term-Forward: On Protecting The Poor Through the Fourteenth Amendment," *Harvard Law Review,* Vol. 83 (Nov. 1969), p. 7.
19. For discussion of minimum protection in the context of health care, see W. T. Blackstone, "On Health Care as a Legal Right: An Exploration of Legal and Moral Grounds," *Georgia Law Review,* Vol. 10 (Winter, 1976), pp. 391–418.

7

ETHICAL CONSIDERATIONS IN CIVIL REMEDIES:

The Equal Employment Context

WILLIAM J. KILBERG

Few aspects of national policy and judicial action have caused more debate than the imposition, by Executive Order or court ruling, of preferences in employment for minorities and/or women. The term "preference" is used here to connote any officially sanctioned pressure upon employers to use racial, ethnic, or sexual criteria in employment decisions.

This essay first summarizes the present state of the law and then questions the equitable considerations which underly much of the judicial reasoning. This essay does not attempt to reach a conclusion. Rather, its aim is to give the reader a more focused view of the issue.

I. A BRIEF STATEMENT OF THE LAW

As a general rule, the courts have shown a dislike for the use of preferences of any kind.[1] Insofar as preferential treatment has been ordered as a remedy for past discrimination, the courts have tended toward the imposition of numerical standards, generally stated as a percentage of hires or referrals allocated to qualified members of a particular racial, sexual, or ethnic category.[2] Although some courts have referred to such preferences as "quotas," in fact no court has ordered the hiring of unqualified persons or demanded that a numerical standard must be met regardless of the availability of qualified persons.[3] Moreover, I am aware of no case where an employer or trade union has been unable to meet a court-ordered numerical standard and raised as a defense to that failure the non-availability of qualified individuals.

It is fair to state that the judiciary has limited the use of preferences to those cases where the factual situation is particularly egregious, setting forth a clear pattern of exclusion of minorities or women, and where the availability of such persons qualified for the particular employment is high.[4]

A careful reading of the case law suggests that the courts have, as a general matter, weighed the impact of preferential relief upon both the members of the discriminated class and those innocent bystanders to the discrimination who might be adversely affected by the remedy. So, for example, where the nature of the discrimination has not been so egregious as to exclude access to employment in a particular job but rather has had the narrower result of limiting promotional opportunities for members of the affected class, the courts have shied away from ordering preferential treatment as part of the required relief.[5]

Where the potential adverse effects of court-ordered preferences are likely to be spread out over a large group, and where the victims of the prior discrimination are themselves easily identifiable, the courts are most likely to order preferential treatment. Conversely, where the victims of the past discrimination are not easily identifiable, but the potential adverse effects of court-ordered preference would fall upon easily identified and innocent third parties, the courts have been loath to provide such relief.

The seniority cases provide a good example of this treatment. Where an affected class has suffered discrimination in job assignment, being placed in less favorable lines of progression and being held there by the use of division or line of progression-based seniority, the courts have not hesitated to provide dramatic relief. Plant-wide or company-wide seniority has been mandated as the appropriate standard whenever an affected class member competes with a non-affected class member. Numerical standards for promotion and transfer have been approved.[6] On the other hand, where minorities and/or women have been laid off in disparate numbers because of the combination of past discrimination and seniority, the courts have been hesitant to order relief which would displace the more senior white males in favor of the less senior members of the affected class.[7] In the latter case, the beneficiaries of any preferential-type relief are persons who, although members of the class previously discriminated against, are generally not themselves clearly identifiable as the subjects of the prior discrimination. The persons who would be displaced, on the other hand, are clearly identifiable. In the former case, however, the persons to benefit by the preferential-type relief are clearly identifiable: they were hired at a time when it can be shown the employer discriminated and were assigned to identified lines of progression on the basis of their race or sex; but the persons who would be adversely affected are those who cannot be easily identified: they are white males who may, at some point in the future, compete for job opportunities with one or more members of the affected class. Furthermore, the white males in the first case are being deprived of no more than an expectation that in the future they would not have to compete against an affected class member using plant-wide or company-wide seniority. That future expectation, moreover, is itself an outgrowth of discrimination; but for the previous discrimination in job assignment, the affected class member would be a competitor. In the second case, however, the white males who are affected by a preferential-type order would lose something much more immediate: their jobs. In both cases, there was a prior discrimination, the effects of which are ongoing. But, because of the equities involved, the first case is an example of relief provided and the second case is an example of relief denied.

The use of the government's procurement authority to further the aim of equal employment opportunity has been sanctioned by the courts. In upholding the legality of the "Philadelphia Plan," by which government contractors are obligated to use all good faith efforts to meet established numerical goals for the employment of minorities, the United States Court of Appeals for the Third Circuit found that such goals could be imposed even absent an express judicial or administrative determination of discrimination. The court declared that such goals are not "a punishment for past misconduct, [t]hey are a covenant for present performance."[8]

As a legal matter, Executive Order 11246,[9] which provides the authority for the imposition of numerical hiring standards on government contractors, is premised on the right and responsibility of the executive branch to determine the terms and conditions upon which it will contract with private parties.[10] It is entirely appropriate for the federal government to use its economic power to assist in the implementation of the national policy of protecting individuals from employment discrimination and remedying the effects of past discrimination.

One of the most pervasive effects of past discrimination is the nationwide underutilization of minorities and women in certain categories of employment. This underutilization is a major target of the Executive Order Program. The regulations issued pursuant to the Executive Order require the development of numerical goals to remedy such underutilization and the setting of timetables within which good faith efforts are to be made to meet the goals.[11] These are encompassed in a written affirmative action program.

To the extent that the affirmative action obligation outlined by the Executive Order regulations is similar to court orders issued under statutory or constitutional authority, it may also be viewed as a preference. Typically, however, the affirmative action program sets forth a timetable longer than that imposed by court order; so, too, the affirmative action program typically does not make reference to specific hiring ratios, as do many court orders premised on a finding of discrimination.

II. THE ETHICAL CONSIDERATIONS

The sensitivity of the courts to the imposition of preferences as a remedy, and the emphasis which the executive branch puts on nondiscrimination in the application of its affirmative action regulations, are both reactions to the reality that the use of preferential-type relief may work an unjust hardship on innocent third parties. Whenever the government threatens the rights or interests of one group or individual in order to provide or preserve the rights or interests of another, our concept of justice requires that a justification be provided that society regards as equitable. Put another way, when we draw the line between conflicting interests, we require convincing equitable reasons for placing the line where we do.

There are three parties at interest: the members of the formerly discriminated group (the disadvantaged class), the members of the non-discriminated group (the non-disadvantaged class), and the employer or prospective employer. There may also be union interests involved, but these are likely to overlap to some degree with each of these other three interests. The degree to which the union interests overlap will depend, *inter alia*, on the degree of minority participation in the union and the effect of any potential relief on the collective bargaining agreement.

A. The Rights of the Disadvantaged Class versus the Rights of the Employer or Prospective Employer

Clearly, any form of employment preference places a limitation on the rights and freedom of the employer. As with requirements placed on employers by all social legislation, as well as economic regulatory legislation, the affirmative action mandate or court-ordered preference limits employer freedom of action. The question that must be asked is whether such a limitation is justified based upon a concept of social justice. Three factors ought to be considered:

1. The responsibility which employers have for the existing employment imbalance which preferential notions seek to rectify. No one would doubt that business shares with other segments of society some responsibility for the initiation or perpetuation of economic patterns which led to and maintain unequal opportunity.

2. The interests of both the employer and the disadvantaged class must be weighed. The impact on the employer—whether defined in economic or social terms—is more diffuse and less a threat to its well-being than the impact on the disadvantaged class of a contrary policy.

3. The employer's interest must be weighed against that of society. Equality of opportunity—realized in fact without extraordinary delay—is a social objective at least as important as those which have been used in the past to justify other forms of regulation. The costs of achieving this objective, insofar as they can be given an economic quantification, are likely to be passed along to the consumer. Society, then, pays the price of its desire for rapid equality, and the employer should not be heard to object.

B. The Rights of the Disadvantaged Class versus the Rights of the Non-disadvantaged

Because in some of its forms it represents a struggle between equally deserving social interests, the balancing between the rights of the disadvantaged class and the non-disadvantaged class is more difficult to deal with than the employer-disadvantaged dichotomy dealt with above. There are a number of questions that must be asked:

1. Is there a violation of legitimate non-disadvantaged class member expectations when an employer searches for and selects a disadvantaged class applicant even though a qualified non-disadvantaged class member is available? Immediacy of availability may be a reasonable concern for the employer, but it hardly grants to an applicant a right to a job or even a moral claim to one. Tougher questions would go to the ability of the employer to discriminate overtly in favor of disadvantaged persons in the advertising or recruitment methodologies used. Would it matter if such techniques drove away potential non-disadvantaged applicants or, put another way, had the effect of excluding non-disadvantaged persons from competition for certain job opportunities?

2. Where two applicants are equally qualified, is there a violation of legitimate non-disadvantaged class member expectations when an em-

ployer is, in effect if not by overt action, compelled by the government to select the member of the disadvantaged class over the equally qualified member of the non-disadvantaged class?

Where two applicants are equally qualified, society allows a variety of personal prejudices of the employer to enter into the selection process. Many of these biases have little or no value to the society as a whole; indeed, they may be harmful to the common good. Certainly, if qualifications are equal, the choice must, by definition, be based on factors with little, if any, job relationship. No one can reasonably expect otherwise. But, what of a government-compelled choice? Is the non-disadvantaged class member being treated equitably when he or she knows that the cards will always be stacked against him or her when competing against an equally qualified disadvantaged class member? Would society tolerate a situation where the non-disadvantaged class member with equal qualifications would always be preferred? Is that not one of the evils the spate of laws and court orders are designed to do away with? Is it crucial to our analysis that no court order and no federal regulation actually requires the selection of the disadvantaged class member? Can one really distinguish a clear command to employ or promote disadvantaged class members in specified numbers from the more typical court-imposed hiring ratios, e.g., one black for every two whites hired? Is it relevant that a goal may have a five-year timetable and that many individual employment decisions will be made so that the difficult decisions become submerged in a mass of numbers? Further, ought we to consider the fact that preferential employment decisions often occur in a context where qualified non-disadvantaged persons, because of past practices unrelated to qualification, have had advantages in selection over equally qualified disadvantaged persons? Finally, should it matter that the competing persons are not themselves either the subjects of advantage or disadvantage but are merely members of a class formerly advantaged or disadvantaged? That, indeed, some members of the disadvantaged class are themselves advantaged (because of superior education or the like), and that some members of the advantaged class are themselves disadvantaged?

3. Is there a violation of legitimate non-disadvantaged class expectations when qualified non-disadvantaged applicants are not selected and disadvantaged applicants requiring training are selected instead? Surely, this is inconsistent with any notions of meritocracy, where the most qualified applicant gets the job. Nevertheless, are there competing social or equitable considerations?

Ought we to consider concepts of relative fault as a justification for selecting the less qualified over the more qualified? One can envision instances where particular non-disadvantaged class members are the direct beneficiaries of previous discrimination and, further, where their greater qualification is itself an outgrowth of their previous unfair advantage. Is it appropriate to "even the score"? Would it make a difference if the beneficiaries of the previous discrimination are not easily identifiable; if the subjects of the previous discrimination are not easily identifiable?

Moving away from notions of relative fault, might we base a preference for a less qualified person on the unfair benefit enjoyed by the non-disadvantaged class as compared to the unfair lack of benefit suffered by the disadvantaged class because of society's previous patterns and practices? Would it make a difference if the previous denial of benefit was based on race, as compared to sex or age? Is one form of discrimination more invidious than another? One might read certain judicial opinions to suggest just that.[12]

III. THE ISSUE IN FOCUS

It is far easier to justify the use of preferences when it is only the interests of the employer and the interests of the disadvantaged class at stake. Where the interests of the disadvantaged must be pitted against those of the non-disadvantaged, the ethical dilemma is great indeed. The judiciary and the administrators have set forth a number of justifications in resolution of the dilemma. As noted in the first part of this essay, these include: the close relationship of the persons to be aided by the relief to the persons harmed by the discrimination; the non-identifiability of the persons who may suffer injury as a result of the relief; the benefit which the persons who may suffer from the relief gained from the previous discrimination; the magnitude and invidiousness of the previous discrimination. All of these are criteria which the courts and government agencies have used in imposing or approving the use of preferences.

Underlying the use of preferences is certainly a belief that the failure to remedy past injustices as quickly as possible would do serious harm to our social fabric; that dramatic action is necessary if we are to avoid the appearance—or the reality—that some classes in our society are permanently disenfranchised. At the same time, there is clearly a realization that preferences are not, in and of themselves, a good thing and that special justifications ought to be demanded for their use. Are these justifications legitimate? Do they assure us that what might otherwise be viewed as inequitable is in fact equitable? Are we accepting an approach to equal employment based not on equity as between the rights of competing individuals, but rather on the equity of the desired result? Are we, in essence, accepting the end as a justification for the means? Given the pervasiveness of discrimination in our society in the past, and given the tremendous subjectivity inherent in most employment decisions, whether invidious factors are at work or not, is this really bad?

NOTES

1. See *United States* v. *Wood Lathers, Local 46,* 471 F. 2d 408 (2d Cir. 1973); *NAACP* v. *Allen,* 493 F. 2d 614 (5th Cir. 1974); *EEOC* v. *American Telephone and Telegraph Co.,* 365 F. Supp. 1105 (E.D. Pa. 1973) *aff'd in part,* 506 F. 2d 735 (3rd Cir. 1974).
2. See, e.g., *Carter* v. *Gallagher,* 452 F. 2d 314 (8th Cir. 1971); *Morrow* v. *Crisler,* 491 F. 2d 614 (5th Cir. 1974); *Rios* v. *Enterpise Ass'n. Steamfitters Local 638 of U.S.,* 501 F. 2d 622 (2nd Cir. 1974).
3. Such a requirement would be unlawful. *Associated General Contractors of Mass., Inc.* v. *Altshuler,* 490 F. 2d 9 (1st Cir. 1973), *cert. denied,* 416 U.S. 957 (1974).
4. *Patterson* v. *American Tobacco Co.,* 535 F. 2d 257 (4th Cir. 1976), *pet. cert. pending* in No. 76-7646; *Crockett* v. *Green,* 10 FEP Cases 165 (E.D. Wis. 1975), *aff'd* 534 F. 2d 715 (7th Cir. 1976).
5. *Bridgeport Guardians, Inc.* v. *Members of the Bridgeport Civil Service Commission,* 482 F. 2d 1333 (2d Cir. 1973); *Kirkland* v. *New York State Department of Correctional Services,* 500 F. 2d 420 (2d Cir. 1975), *rehearing en banc denied,* 531 F. 2d 5 (2d Cir. 1975).
6. *Quarles* v. *Philip Morris, Inc.,* 279 F. Supp. 505 (D.C, Va. 1968); *U.S.* v. *Bethlehem Steel,* 446 F. 2d 652 (2d Cir. 1971). *In re Bethlehem Steel Corp.* (Sparrows Point), Decision of the Secretary of Labor, Docket No. 102-68 (Jan. 15, 1973); *U.S.* v. *Allegheny Ludlum Industries, Inc., et al.,* 517 F. 2d 826 (5th Cir. 1975), *cert. denied,* 44 U.S.L.W. 3593 (1976); *EEOC, et al.,* v. *AT&T,* 12 EPD ¶11160, 13 FEP 392 (E.D. Pa. 1976 not otherwise reported), app. pndg.
7. *Watkins* v. *United Steelworkers of America, Local 2369,* 516 F. 2d 41 (5th Cir. 1975). *Waters* v. *Wisconsin Steel Works of International Harvester Co.,* 502 F. 2d 1309 (7th Cir. 1974), *cert. denied,* 96 S. Ct. 2214 (1976). *Jersey Central Power & Light Co.* v. *Local Union 327, etc. of IBEW,* 508 F. 2d 687 (3rd Cir. 1975), judgment vacated and remanded *sub nom EEOC* v. *Jersey Central Power & Light,* 96 S. Ct. 2196 (1976).
8. *Contractors Association of Eastern Pa.* v. *Shultz,* 442 F. 2d 159, 176 (3rd Cir. 1971), *cert. denied,* 404 U.S. 851 (1971).
9. 3 C.F.R. 174 (1973).

10. *Contractors Association of Eastern Pa.* v. *Shultz,* supra.; *Perkins* v. *Lukers Steel Co.,* 310 U.S. 113 (1940).
11. 41 C.F.R. 60-2.10, *et seq.*
12. Compare *Weeks* v. *Southern Bell Telephone & Telegraph Co.,* 408 F. 2d 228 (5th Cir. 1969) with *Usery* v. *Tamiami Trail Tours,* 531 F. 2d 224 (5th Cir. 1976), aff'q sub nom. *Hodgson* v. *Tamiami Trail Tours,* 4 F.E.P. 728, 4 E.P.D. ¶7795 (S.D. Fla. 1972).

IV

ETHICAL CONSIDERATIONS IN PUBLIC SERVICE STRIKES

8

PUBLIC SECTOR STRIKES:

Legal, Ethical, and Practical Considerations*

JOHN F. BURTON, Jr.

The emergence of strong public sector unions is one of the most significant developments of the past decade. This development is not only important for the wages and working conditions of government workers, but is also interrelated with several critical topics of recent years, including the financial plight of large cities. Another consequence of the growth of public sector unions is the necessity to reexamine much of the conventional wisdom about industrial relations. Indeed, the new views of industrial relations emerging from the public sector may soon force a revision of private sector collective bargaining.

I. HOW DECISIONS CONCERNING WORKING CONDITIONS ARE MADE

Decisions concerning working conditions must be made in every employment relationship. Working conditions include wages, fringe benefits, hours, the nature of the job task, the amount of job security, the

* Much of the work on this essay was done as part of the *Studies of Unionism in Government*, which are being conducted by the Brookings Institution with financial support from the Ford Foundation. The views are the author's and not those of the officers, trustees, or staff members of the Brookings Institution or of the Ford Foundation.

Helpful comments on a preliminary draft of the essay were received from many persons, including the participants in the University of Kansas Symposium on Ethics in Business and the Professions; Melvin Reder, George Neumann, Wes Wildman, Bernard Meltzer, and Joel Seidman of the University of Chicago; and Paul Gerhart, Charles Rehmus, and Ted Clark. All disagree with at least some of my analysis.

procedures and standards for promotions and discipline, and a myriad of related topics.

The methods for making the decisions about working conditions will vary between the public and private sectors and between unionized and unorganized employees. However, certain factors affect all employers, albeit with varying intensity. One factor is supply and demand forces in the labor market, which largely determine working conditions. An employer offering unattractive working conditions will have difficulty attracting workers, while profligate working conditions translate into prices that repulse customers, thereby reducing demand for workers. Another factor affecting all employers is legal constraints for certain aspects of working conditions. Wage differentials based on race or a grossly unsafe environment are examples of working conditions that cannot legally be offered even if some workers would accept them.

Private and public sector employers may respond differently to similar labor market conditions and legal constraints. In the absence of a collective bargaining relationship, a private sector employer can offer those working conditions that maximize profit. The goal of profit maximization may be something of a simplification, but the private sector employer's criterion for decision making is relatively straightforward compared to the criterion in the public sector. There, profit maximization may operationally be virtually impossible since explicit prices are not charged for much of the output. Nor is profit maximization an adequate conceptual basis for understanding the motivation of public officials; alternative models suggest that decisions on employment and other matters will be related to political considerations, such as getting reelected. In addition to having a criterion for deciding issues that is conceptually and operationally less apparent, the public sector employer typically is more fettered in reaching decisions on working conditions than the private sector employer. Many governmental units divide authority between the nominal employer and a civil service commission. Also, laws are common which mandate how certain working conditions shall be established. Ordinances which require municipalities to pay the "prevailing wage" in the private sector are an example of limits on the discretion of public sector employers that have few counterparts in the private sector.[1]

Decisions about working conditions will also be made differently in unionized and non-unionized firms. The collective bargaining contract limits the potential scope of discretion for management, and the actual decisions by employers are subject to challenge and reversal through the grievance-arbitration procedure found in most contracts.

There are several reasons offered to justify unions and collective bargaining.[2] The first is that unions have desirable economic consequences. Unions can offset the monopsony power of employers, which in the absence of unions would cause wages to be below the competitive level. A related assertion is that unions may overcome deficiencies in labor market information and mobility. A second reason is that unions and collective bargaining have desirable non-economic consequences for workers. Although collective bargaining contracts limit employer discretion, the employees gain greater job security and protection from arbitrary treatment. This protection is underwritten by the availability of the grievance-arbitration procedure. Third, collective bargaining leads to industrial peace because the potential discontent of workers over onerous conditions is assuaged by their participation in the decision-making process. Fourth, unions provide workers with a means of effectively participating in the political arena. A final reason is that the combination of the previous advantages of unicns and collective bargaining gives workers a sense of well-being that makes them defenders of the basic economic system and unsympathetic to radical reforms.

There are also several reasons offered to disparage unions and collective bargaining. First, undesirable economic consequences can result from union distortion of the wage structure or from direct limits on the efficient utilization of labor, both of which result in misallocation of resources. Second, union security arrangements and emphasis on seniority encourage mediocrity among workers and discourage initiative. Third, the emergence of unions almost inevitably results in overt manifestations of industrial unrest, such as strikes and lockouts.

These are some of the arguments offered to justify or condemn unions and collective bargaining. Each argument has been criticized on theoretical and empirical grounds, and some are suspect. However, most industrial relations scholars support unions and collective bargaining because the arguments in favor appear to outweigh the negative assertions.[3] Likewise, our public policy at the federal level and—to a lesser degree—at the state level also encourages unions and bargaining, presumably because the arguments favoring unions are more persuasive. The balance of this essay assumes that our public policy should generally encourage collective bargaining when employees desire to use this approach, and examines several issues that flow from this assumption.

II. THE ROLE OF STRIKES AND OTHER FORMS OF CONCERTED ACTIVITY

The endorsement of collective bargaining and trade unions is not equivalent to the endorsement of strikes. However, most supporters of unions and bargaining also believe the strike over economic disputes should be legal, at least in the private sector. Even in the private sector, however, strikes over organizing disputes and over grievances about existing contracts are discouraged.

Why is the Strike Model of collective bargaining generally preferred to the No-Strike Model? Presumably because most of the advantages of collective bargaining previously described are enhanced when the strike is legal. The strike appears generally to increase the bargaining power of unions, thereby increasing the beneficial economic and non-economic consequences of collective bargaining. Likewise, the possibility that a strike will impose costs on the parties makes them bargain seriously rather than allowing disagreements to fester, and thus the availability of the strike promotes industrial peace. In addition, the strike provides a relatively acceptable outlet for worker discontent compared to alternatives such as sabotage.

These alleged virtues of the Strike Model are viewed by some with skepticism, although the net advantage of the Strike Model over the No-Strike Model is generally accepted for the private sector. However, for the public sector, prevailing sentiment favors the No-Strike Model.[4] The justification for a differential treatment of the public and private sectors contains a number of aspects, which I believe can be reduced to two themes. One is that the economic restraints from the labor and product markets, which provide a check on union power, are less compelling in the public sector than in the private. The other is that the nature of the political process makes public sector managers less able than private sector managers to resist union demands that are coupled with the threat of a strike. These assertions about the differences between the public and private sectors are traditionally made without much, if any, supporting evidence, and in the absence of the evidence, I find the assertions unconvincing. Some of the evidence relevant to the assertions is reviewed in later sections, and the public sector–private sector distinctions for strikes will be examined in more detail.

Putting aside for now the arguments that would distinguish the public and private sectors, the differences between the Strike Model of collective bargaining and the No-Strike Model deserve further scrutiny. One reason the Strike Model works is that, when the employer and trade union disagree, costs are imposed on the parties and later, if not sooner, the parties reach an agreement because the costs of continuing the fray exceed the possible gains. If all of the costs of strikes fell on the parties to the negotiations, there would be little concern by others about the process. However, consumers, suppliers, creditors, and taxpayers all find themselves bearing some of the costs of strikes, and these costs for non-participants can become the determining factors for public policy. Indeed, one reason why national emergency strikes in the private sector and most strikes in the public sector receive differential treatment in public policy is that the ratio of the non-participants' cost to the participants' cost may be higher for these strikes than for ordinary private sector strikes. In the public sector, for example, some employers are hardly affected by the strike in the conventional sense of suffering lost revenues; tax receipts may accumulate while the strike proceeds. Under these circumstances, and especially if government services are curtailed by the strike, the public may have to bear an abnormally high share of the costs of the strike.

Consideration of the costs of disagreement and how the costs are divided between the participants and non-participants in a dispute should not be confined to the strike, because there are numerous other forms of concerted activity used by employers and unions. Picketing, boycotts, and violence have a long if not venerable history in labor relations. Less obvious techniques include work slowdowns, sabotage, blacklisting, and—more prevalent in the public sector—involvement in political activity, including participation in a patronage system.[5] These forms of gaining influence over working conditions vary in the nature of the costs imposed on management and labor, in the visibility of the costs, and in the ratio of the costs imposed on participants in the bargaining process compared to the cost for non-participants. A full assessment of the Strike Model and the No-Strike Model of collective bargaining must consider the likely changes in the use of these other forms of concerted activity if the strike is encouraged or discouraged.

III. PROCEDURES TO REDUCE THE USE OF STRIKES

Strikes and other forms of concerted activity are usually not considered desirable per se but a cost necessary to make collective bargaining work. The undesirable aspects of strikes and other forms of concerted activity provide one reason why society limits the parties' abilities to use these weapons and why society interferes in the process used by unions and management to decide working conditions. There are, however, several other reasons why public policy may interfere in the decision-making process in addition to the interest in reducing the amount of conflict between the parties. A second reason is that the public policy is sometimes designed to change the relative bargaining power of the parties. Weak unions or weak employers may find their bargaining power enhanced by government policies. Third, the government may become involved in private disputes in order to improve the quality of the outcome. Employers and unions may find that government expertise interjected into a dispute leads to better solutions than the parties would have reached on their own. Fourth, the government may intervene to protect the public against the employer and the union. For example, public policy may outlaw contracts with extensive limits on competition that are beneficial to the parties but impose undue costs on the public.

Society's intervention into the collective bargaining process involves two approaches (often coupled): (1) limitations or prohibitions on certain forms of concerted activity, and (2) the provision of impasse resolution procedures that are substitutes for concerted activity.

Limitations or Prohibitions on Concerted Activity. Certain types of concerted activities are prohibited because they impose excessive costs on one of the parties, thus increasing in an unacceptable manner the bargaining power of the other party. Thus, blacklisting, violence, and certain forms of picketing and boycotts are prohibited. Other types of concerted activity are impermissible because of the actual or alleged excessive harm they impose on non-participants. Fines for public sector strikers at least partially reflect this concern. Concerted activity is sometimes limited because of similar concerns about the impact on the parties or non-participants. The temporary injunctions for national emergency disputes and the restricted duration for recognition picketing are examples of limits on concerted activity.

Substitutes for Concerted Activity. Public policy often encourages and sometimes compels the use of impasse-resolution procedures that are substitutes for concerted activity.

Mediation involves a third party who attempts to bring the disputants together by persuasion. Sometimes the mediator convinces the parties that a compromise is desirable or that a solution the mediator introduces is acceptable. The mediator has no power to force a settlement, and the mediator's suggestions normally are unpublicized. The successful mediator can reduce the extent of conflict and sometimes improve the quality of the bargaining outcome, but ordinarily cannot change the relative bargaining power of the parties. Mediation is the third-party procedure most widely used to reduce disputes about the terms to be included in a collective bargaining contract. The Federal Mediation and Conciliation Service provides mediators for private sector negotiations and for government negotiations in those states without state mediation services.

Fact finding involves a third party who investigates a dispute and issues a report that indicates the causes of the dispute and that usually contains recommendations for settlement. The recommendations are not binding on the parties, although the fact finder's report usually is made public in order to generate pressure for acceptance of his recommendations. Probably the major purpose of fact finding is to reduce conflict, although the procedure may affect the relative bargaining power of the employer and the union. In addition, the fact finder's opinion may introduce elements of public interest into the dispute that the parties would otherwise ignore. This can occur, for example, when the fact finding is mandated by a statute that included criteria for the fact finder's recommendation. The ability of a fact finder to consider criteria that otherwise would be ignored by the parties is limited, however, because the selection of a fact finder is usually at least partially controlled by the parties, and the fact finder who vigorously crusades for the public interest will probably have a brief, if eventful, career. Fact finding is voluntary in the private sector, and is rarely used, especially outside the railroad industry.[6] A number of states have statutes that permit or require fact finding in disputes involving state or local government employees.

Arbitration involves a third party who investigates a dispute and issues a decision with recommendations that are binding on the parties. The most common form of arbitration is used to resolve disputes about the meaning of a collective bargaining contract that the parties have already signed. This type of arbitration, termed arbitration of rights, is the final step in a grievance procedure. In the private sector, arbitration of rights, although not required by statute, is present in almost all collective bargaining contracts, usually in conjunction with a union pledge that strikes will not occur during the life of the contract. Decisions by the National Labor Relations Board and the federal courts have encouraged the use of this type of arbitration, often based on the argument that the result will be fewer strikes. Grievance arbitration is less prevalent in the public than in the private sector.[7] In some states, this reflects legal restrictions prohibiting delegation of governmental authority to third parties. Also, since arbitration of rights is often the *quid pro quo* for a no-strike clause, and since strikes are generally illegal in the public sector, public sector unions have less to offer management in exchange for an arbitration provision.

The major purpose of arbitration of rights is the reduction of conflict. Since the arbitrator is chosen by the parties to interpret an existing contract, the procedure should not change the relative bargaining power of the parties. Nor is this type of arbitration able to reflect the public's interest when that is in conflict with the interests of labor and management.[8]

Another form of arbitration, known as arbitration of interests, is used to resolve disputes about the terms to be included in a collective bargaining contract. This type of arbitration is not required by any statute applicable to the private sector. In a few instances, private sector employers and unions have voluntarily agreed to submit unresolved bargaining issues to arbitration. Arbitration of interests is also unusual in the public sector, although a few states now require arbitration when disputes for police and firemen cannot otherwise be resolved, and Iowa requires arbitration for all unresolved bargaining disputes. In addition, several states allow, but do not require, government units and unions to submit unresolved disputes to arbitration.

Arbitration of interests can change the relative bargaining power of the participants, since the arbitrator sometimes mandates a solution that a party could not otherwise achieve. Also, the arbitrator can represent the public interest vis-à-vis the interests of the parties. The result may occur if the arbitration is mandated by a statute that includes criteria for settlement. Moreover, the arbitrator can make decisions superior to the outcomes the parties would have reached on their own—or inferior, perhaps reflecting the arbitrator's lack of familiarity with the details of the employment relationship or the arbitrator's unwillingness to make a decision that one party would find particularly repugnant, such as a change in staffing requirements that reduces employment opportunities. These possible consequences of arbitration help explain the general reluctance of labor and management to use arbitration or to support the enactment of statutes encouraging or mandating arbitration.

Another reason for the intervention of public policy into the collective bargaining process is to reduce conflict and promote voluntary settlements, but it is possible that arbitration of interests has the opposite effect. If the parties expect that the arbitrator's recommendation will be a compromise between the parties' final positions, then the union and employer have an incentive to exaggerate their stands. This centrifugal force will reduce the chances that the parties will reach a settlement on their own, in contrast to the ordinary process of collective bargaining, in which the parties converge to a solution. In order to overcome this defect of ordinary arbitration, the arbitrator can be required to choose one of the parties' final offers. Final-offer arbitration means the arbitrator's decision cannot be a compromise between the parties' last offers. In theory, this form of arbitration induces each party to produce the most reasonable final offer in order to induce the arbitrator to select that position, and as the parties rush toward reasonable positions, they reach a settlement and eliminate the need for arbitration.

Impact of Efforts to Reduce Concerted Activity. Have the public policies that limit or prohibit concerted activity or provide substitutes had an impact? The type of concerted activity most readily quantifiable is the work-stoppage (strikes and lockouts), and some data relevant to the issue of public policy impact are presented in Tables 1 and 2.

Several dimensions of strike activity can be traced. Published statistics pertaining to the entire economy, reproduced in Table 1, are the number of work stoppages beginning in the year (column 1), the number of workers involved in strikes beginning in the year (column 3), and the number of man-days idle during the year (column 5). In order to facilitate comparisons through time and between the entire economy and the government sector, these measures of strike activity have been related to the number of full-time equivalent (FTE) employees (column 8). The results are the number of strikes per one million FTE employees (column 2), the number of striking employees per one thousand FTE employees (column 4), and the number of man-days idle per one thousand FTE employees (column 5). The mean duration of strikes (column 7) has been calculated by determining the number of man-days idle during the year per striking worker. Table 2 presents the equivalent data for the state and local government sector.

The per-employee data in Table 1 indicate several broad swings in strike activity in recent decades. From a low base in the first half of the 1930's, there was a substantial increase in strike activity following the enactment of the National Labor Relations Act and the surge in union membership. Another increase in strike activity occurred in the few years following World War II. Measured by the number of workers involved and man-days idle (both per thousand FTE employees), the 1945–49 record represents a peak. This was followed by about fifteen years of a general decline in strike activity, with the 1960–64 period standing as the low point of the post-war period. The level of strike activity since 1965 is clearly above the record of the early 1960's, a surprising result since the share of the non-agricultural labor force organized has continued to decline during this period. (Currently

TABLE 1. Work Stoppages in the Entire Economy
Annual Averages, 1930–75

Years	*Work Stoppages* Number (1)	per million FTE Employees (2)	*Workers Involved* Number (thousands) (3)	per thousand FTE Employees (4)	*Man-days Idle* Number (thousands) (5)	per thousand FTE Employees (6)	*Duration* Mean in days (7)	*FTE Employees* (thousands) (8)
1930–34	1168	39.2	698	23.4	11442	384.0	16.4	29798
1935–39	2862	82.2	1125	32.3	16950	486.8	15.1	34818
1935	2014	62.9	1120	35.0	15500	484.4	13.8	32000
1936	2172	61.8	789	22.5	13900	395.8	17.6	35123
1937	4740	130.4	1860	51.2	28400	781.3	15.3	36349
1938	2772	80.1	688	19.9	9150	264.3	13.3	34619
1939	2613	72.6	1170	32.5	17800	494.5	15.2	35997
1940–44	3694	87.0	1575	37.1	11220	264.4	7.1	42440
1945–49	4091	90.9	3046	67.7	54640	1213.7	17.9	45020
1946	4985	113.2	4600	104.4	116000	2633.6	25.2	44047
1950–54	4651	92.9	2420	48.4	34340	686.1	14.2	50052
1955–59	3844	72.6	1976	37.3	34140	644.9	17.3	52941
1960–64	3466	62.1	1316	23.6	18600	333.2	14.1	55823
1965–69	4742	73.6	2302	35.7	36537	566.9	15.9	64446
1970	5716	83.5	3305	48.3	66414	969.7	20.1	68941
1971	5138	75.1	3280	47.9	47589	695.3	14.5	68444
1972	5010	71.8	1714	24.6	27066	387.8	15.8	69795

1973	5353	73.3	2251	30.8	27948	382.6	12.4	73045
1974	6074	82.0	2778	37.5	47991	647.9	17.3	74071
1975	5031	70.1	1746	24.3	31237	435.2	17.9	71769

Sources: 1930–73 data on work stoppages, workers involved, and man-days idle are from U.S. Department of Labor, Bureau of Labor Statistics, Bulletin 1877, *Analysis of Work Stoppages 1973* (Washington: G.P.O., 1975). 1974–75 data are from Bureau of Labor Statistics, Summary 76-7, *Work Stoppages 1975* (August 1976), Table 1, p. 3. Data on full-time equivalent (FTE) employees are from U.S. Department of Commerce, Office of Business Economics, *The National Income and Produce Accounts of the United States, 1929–65, Statistical Tables,* Table 6.4 (Washington: G.P.O., 1966), and from Table 6.4 (Table 6.7 for 1976) of the July issues of the *Survey of Current Business* for 1968–76.

Note: Duration is the number man-days idle (Col. 5) divided by the number of workers involved (Col. 3).

TABLE 2. Work Stoppages in the State and Local Sector Annual Averages, 1945–74

	Work Stoppages		*Workers Involved*		*Man-days Idle*			
Years	Number (1)	per million FTE Employees (2)	Number (thousands) (3)	per thousand FTE Employees (4)	Number (thousands) (5)	per thousand FTE Employees (6)	*Duration* Mean in days (7)	*FTE Employees* (thousands) (8)
1945–49	28	8.7	3.7	1.1	19.5	6.1	5.3	3207
1950–54	31	7.7	5.0	1.3	31.7	8.0	6.3	3971
1955–59	19	3.9	1.9	.4	17.8	3.6	9.4	4879
1960–64	31	5.2	18.0	3.0	52.1	8.7	2.9	5987
1965	42	6.1	11.9	1.7	146.3	21.3	12.3	6878
1966	142	19.7	105.1	14.6	455.0	63.0	4.3	7221
1967	181	24.1	131.7	17.6	1246.3	166.2	9.5	7498
1968	251	32.0	200.2	25.5	2535.6	323.0	12.7	7850
1969	409	50.2	159.5	19.6	744.6	91.4	4.7	8150
1970	409	48.1	177.7	20.9	1375.1	161.8	7.7	8501
1971	327	36.9	151.6	17.1	893.4	100.9	5.9	8854
1972	375	40.5	142.1	15.4	1257.2	135.9	8.8	9253
1973	386	40.2	196.0	20.4	2299.3	239.2	11.7	9613
1974	382	38.6	160.1	16.2	1402.7	141.6	8.8	9905

Sources: Data on work stoppages, workers involved, and man-days idle are from U.S. Department of Labor, Bureau of Labor Statistics, "Work Stoppages in Government, 1974," Report No. 453 (1976). Source of data on full-time equivalent employees is cited in Table 1.

Note: Duration is the number of man-days idle (Col. 5) divided by the number of workers involved (Col. 3).

about a quarter of all nonfarm workers are organized, compared to about a third of such workers in the early 1950's.)

The data in Table 2 for state and local government employees present an even more dramatic story of increased strike activity after 1965. It is interesting that the public sector strike surge took place in a few years in the late 1960's and that since 1970 there has been no further increase in the level of strike activity. Comparison of Table 2 with Table 1 indicates that the state and local sector has on the average much less strike activity on a per employee basis than the rest of the economy.

What can be said about the factors that explain the level of strike activity? In a recent study that I coauthored with Charles Krider (Burton and Krider, 1975), the previous empirical examinations of strike activity were reviewed. We concluded (on pp. 148–49) that

> . . . our empirical knowledge of the public and private sectors is both narrow and fragile. Narrow because most examinations have concentrated on the number of strikes. Other measures of strike activity, such as the number of striking employees and the number of man-days lost, have received less attention, and the duration of strikes has been virtually ignored. Our knowledge can also be characterized as narrow because most studies have concentrated on national time-series data; studies of disaggregated or cross-sectional behavior are uncommon. The evidence is fragile because the results are sensitive to the specific time period or level of aggregation from which data are used.

In short, most previous studies of strike activity were unimpressive in terms of their predictive ability.

Krider and I then attempted to develop a model that would explain the variations among states in the number of strikes by government employees in the local noneducation sector. We included a number of variables pertaining to each state's work force and economic conditions, such as the relative earnings of public employees compared to private sector earnings and the state's unemployment rate. In addition, we examined the relationship between the amount of strike activity in a state and a number of public policy variables, such as the availability of mediation or fact finding, the presence of statutory penalties for strikers, and the presence or absence of a statute encouraging or discouraging collective bargaining by local government employees. Our empirical evidence suggested that laws encouraging collective bargaining perhaps have a mild tendency to encourage strikes, but that the other public policy variables have no demonstrable impact. In particular, the availability of fact finding or mediation had no relationship with the amount of strike activity in the state. Of most relevance for the present paper, the Burton and Krider study, after considering the new evidence on public sector strikes as well as the evidence from earlier studies of both the private and public sectors, indicated (on p. 172) that "the industrial relations research community and policy makers have almost no evidence on which to base assertions about the impact of public policy on strike activity."

An important qualification should be added to the Burton and Krider evidence. Our data pertained to 1968–71, when fact finding and mediation were the only types of third-party procedures generally available. More recent evidence pertaining to arbitration statutes for police and fire fighters should provide some encouragement to policy makers seeking third-party procedures that effectively reduce the amount of strike activity.[9]

Other Consequences of the Use of Substitutes for Sanctions. The preceding paragraphs indicated that mediation and fact finding have no demonstrable impact on the amount of strike activity, and that arbitration may reduce the number of strikes. These procedures also have other consequences, and in recent years a few studies of these consequences have appeared.

One critical issue is whether the third-party procedures produce wage settlements that are significantly different than the wage agreements the parties would have reached on their own. A study of final-offer arbitration for police and fire fighters in Michigan and Wisconsin concluded that the initial impact of the procedure was to raise salaries by more than 1 but less than 5 percent and that in subsequent years the effect was smaller or non-existent (Stern, 1975, p. 169). The same study compared conventional arbitration for police and fire fighters (used in Pennsylvania) with final-offer arbitration (used in Wisconsin and Michigan) and found no difference in the effect on wages (Stern, 1975, p. 166). A New York law in effect since 1974 uses conventional arbitration to resolve disputes involving police and fire fighters. A recent study (Kochan, 1976, p. 3) estimated that the law resulted in wages 0 to 2 percent higher than the wages that would have been expected under the pre-1974 statute, which relied on fact finding. These studies, while limited in scope, suggest that use of third-party procedures does not have a significant impact on the economic consequences of collective bargaining.

Another important issue pertaining to the use of procedures such as arbitration is whether their availability undermines the collective bargaining process by making labor and management dependent on a third party to reach an agreement. A recent survey compared the extent of reliance on the strike in private sector negotiations (about 15 percent of major negotiations ended in strikes) with the reliance on third-party procedures (other than mediation) when they were available. Conventional arbitration was more likely to be used than strikes, but the level of final-offer awards, fact-finding reports, and strikes were at roughly the same level (Feuille, 1975). The study of the New York statute providing for arbitration of disputes involving police and fire fighters found that, while there was a 16 percent increase in the probability of an impasse occurring in negotiations, there was no significant change in the amount of compromising behavior of the parties, and that bargaining was no more inhibited under arbitration than under fact finding (Kochan, 1976, p. 3). Again, the evidence is sketchy, but third-party procedures apparently do not have an inevitable chilling effect on collective bargaining.

The use of third-party procedures has been criticized because the quality of the arbitrator's decisions may be deficient. In grievance arbitration, the typical issue is legal in nature because it involves the interpretation of contract language. In contract arbitration, however, the arbitrator must write the contract and the issues often involve economics, such as the level of wages and fringe benefits. Horton (1975) and Berkowitz (1976) argue that since most arbitrators are lawyers, not economists, they may lack the training, expertise, or attitude necessary to make appropriate decisions on these issues.

This review of the consequences of utilizing substitutes for sanctions, such as arbitration, cannot be conclusive because of the insufficient evidence now available. The fragmentary evidence suggests that, in comparison to a model of collective bargaining that relies on strikes as the motivating force, a model that relies on third-party procedures to resolve impasses is neither a panacea nor a catastrophe.

IV. THE IMPACT OF UNIONS AND COLLECTIVE BARGAINING

There are numerous studies of the impact of collective bargaining and unions. Most have concentrated on the economic consequences, such as increases in wages, but non-economic consequences have also been examined.

The union impact on relative wages in the private sector has been extensively evaluated. The classic study by H. G. Lewis (1963) estimated that the average effect of all American unions on the wages of their members relative to non-union members was between 10 and 15 percent. More recent studies of wages in the private sector have fallen into two camps. Some studies continue to show a union impact of about 10 to 15 percent (Boskin, 1972). Other studies have found a much larger impact of unions on relative wages (Ryscavage, 1974). A possible reconciliation of the different estimates is that some of the studies showing the higher figures are based upon data from the 1970's, while the lower estimates are primarily derived from examination of data from the 1960's. There is evidence indicating that union wage gains have outpaced those for non-union workers in the last ten years (Scheible, 1975, and Ashenfelter, 1976).

We are unlikely ever to reach total agreement about the impact of unions on relative wages because it is difficult to disentangle the influence of unions from other factors that influence wages. Inter-industry wage differentials, for example, can be affected not only by unions but by other economic factors, such as industry profitability, the degree to which industry output is concentrated in a few sellers, and the growth of industry employment and productivity.

The presence of a variety of factors influencing wages also complicates the task of determining the impact of unions on public sector wages. Evidence is now accumulating that public sector workers generally earn more than private sector workers with equivalent training and in equivalent occupations.[10] This generalization must not obscure the relatively low compensation of public sector workers compared to private sector workers in many states and localities. However, assuming that public sector workers generally have wages that equal or exceed those in the private sector, to what extent is this due to public sector unions and collective bargaining?

David Lewin has recently completed a survey of the studies of the influence of public sector unions on wages that provides evidence relevant to this question. Lewin (1977) reports that

> The "average" wage effect of unionism in government, according to these studies, is roughly on the order of five per cent, a much smaller impact than is popularly supposed and smaller than the average union wage impact in private industry.

It is necessary to reconcile the finding that public sector wages in general are higher than private sector wages with the Lewin conclusion that public sector unions have a relatively limited impact on wages compared to the impact of private sector unions.[11] An obvious possibility is that factors other than unions explain the high public sector wages; perhaps the prevailing wage procedures, or the decision-making process used by public officials to determine wages, or the public's concern for high-quality public servants are the essential factors. Another possibility is that public sector unions somehow, through economic, political, or legal pressures, boost the wages of all public sector workers compared to private sector workers. Most of the empirical studies compare wages of unionized public sector workers to those of non-unionized public sector workers; such studies would underestimate the union influence on wages if that influence had operated to the benefit of all public sector workers. Thus, it is possible that public sector unions have a greater impact on wages than is suggested in the articles surveyed by Lewin. Nonetheless, the most reasonable conclusion that can be derived from Lewin's survey is that public sector unions apparently have less influence on wages than do private sector unions.

The economic consequences of trade unions are not confined to their impact on relative wages. Indeed, in the private sector, the greatest impact of unions on resource allocation appears to result not from the distortion of the wage structure but from the union's direct restrictions on output through control of manning requirements, the work pace, and work practices (Rees, 1963). I am unaware of a comparable national estimate of the misallocation of resources in the public sector resulting from unions' restrictions of output; so, a comparison of private and public sector unions on this count is impossible.

Another economic consequence of unions is their impact on fringe benefits, such as pensions. For public employees, particularly those in the federal government, some assert that the value of their pensions exceeds those typical in the private sector. But whether the generosity of public sector pensions is due to unions is unclear and, in any case, there is no evidence that public sector unions have more influence than private sector unions on pensions or other non-wage economic matters.

Private and public sector trade unions can also be compared in terms of their non-economic consequences, although these comparisons are also elusive. The basic study that surveys the non-economic consequences of private sector unions is the volume by Slichter, Healy, and Livernash (1960). They documented the narrowing scope of management discretion, the development of management by policy, the changes in management structure, and other results of collective bargaining. There is no study of comparable scope of the impact of public sector bargaining; among the themes that are emerging in partial studies of the sector are the erosion of the civil service system as bargaining emerges as an alternative procedure to determine working conditions, and the centralization of management authority, in contrast to the fragmented management structure typically found in non-union governmental units (Burton, 1972). Opinions differ about the desirability of these public sector developments, but they seem no more significant than the non-economic consequences of private sector bargaining.

This section has reviewed some of the economic and non-economic consequences of public sector unions and private sector unions. For the comparisons where the evidence is most reliable, namely, the impact of unions on relative wages, the results suggest that public sector unions have less influence than private sector unions. For most other aspects of the comparison, it is difficult if not impossible to draw conclusions about the relative impact of public and private sector unions. Under these circumstances, skepticism seems to be the appropriate response to any policy based on an assertion that unions in the public sector have a greater potential or actual influence than private sector unions.

V. HOW SHOULD COLLECTIVE BARGAINING IMPASSES BE RESOLVED?

The previous sections reviewed a number of topics, including: (1) the methods used to decide working conditions in unionized and non-unionized organizations, (2) the role of strikes in collective bargaining, (3) the sanctions other than strikes that parties use to increase their bargaining power, (4) the procedures, such as mediation, which are designed to reduce the utilization of sanctions, and (5) the economic and non-economic impacts of unions. Throughout the survey of these topics, the public and private sectors have been compared and contrasted.

With this background, I will now consider the question: How should collective bargaining impasses be resolved? In particular, the issue of legalizing the strike for public employees will be reviewed.

I believe we are in a new phase of the arguments concerning the appropriateness of strikes for public sector employees. The first phase was initiated in 1966 when the Taylor Committee Report was submitted in New York, sparking a debate characterized by the exchange between Wellington and Winter, on one hand, and Burton and Krider, on the other, that took place in the *Yale Law Journal* in 1969–70. The Wellington-Winter position was that public sector strikes are inappropriate, a conclusion largely derived from a series of logical propositions about why the private sector and the public sector differ. The Burton and Krider response was that the Wellington-Winter arguments could be matched by contrasting logical assertions, and that no supporting evidence was provided for the asserted differences between the public and private sectors. Krider and I provided some data suggesting that the differences between the private and public sectors were more illusory than real. We concluded that Wellington and Winter had not established a significant difference between the public and private sectors, and therefore had not justified differential strike policies. Moreover, we argued that Wellington and Winter had not provided any meaningful distinction between strikes and other forms of collective action used by public employees, such as lobbying, and therefore could not justify a policy that outlawed strikes but not the other activities. We argued that strikes for local government employees should be legalized except for police, fire fighters, and other essential workers. In essence, we argued that the relevant distinction was not private versus public sector workers, but essential versus other workers.

The second phase of the debate on public sector strikes is under way. The fundamental difference between the current arguments and the earlier exchanges is that we now have more evidence about the impact of public sector bargaining. Some of this evidence was surveyed in the previous section, which concluded that there is no evidence that on the average, public sector trade unions have a greater impact on wages or other matters than private sector unions. I believe this reinforces the conclusions in the Burton and Krider article (1970), namely, that different policies for strikes in the public and private sectors seem unwarranted, and that the burden of justifying differential policies should be placed on those who draw the distinction.[12]

While the debate among industrial relations scholars continued during the 1970's concerning the desirability of legalizing the strike for public sector employees, interesting developments occurred at the political level. In 1970, strikes for any public sector employees were legal in only two states (Burton and Krider, 1970, pp. 435–36). By 1975, eight states had legalized public sector strikes (Lewin, 1976). There are, to be sure, a number of limitations placed on these grants of the right to strike, such as requirements that mediation or fact-finding procedures be exhausted before the strikes can occur. Despite these limitations, I believe the movement toward legalization of public sector strikes has been impressive and surprisingly rapid.

The Ethics of Public Sector Strikes. For the purpose of this essay, ethics is defined as the evaluation of conduct on the basis of a recognized standard of behavior.[13] Conduct that does not comply with the behavioral standards is unethical.

There are several sources for the standards of behavior, including religious doctrine, constitutional law, the "higher law,"[14] statutory law, professional codes of conduct, and group mores. This list of sources is not meant to be inclusive, but is sufficiently diverse to raise an obvious question: How is conduct to be evaluated when it can be evaluated by more than one standard and they are in conflict?

Some conflicts between behavioral standards can be "resolved" with relative ease. An act banned by a state law may be protected by the U.S. Constitution, and thus, at least in an operational sense the ethical dispute is resolved. But many conflicts between the standards used to evaluate conduct are not so easily resolved. For example, efforts to amend the Constitution in order to outlaw certain conduct often provoke arguments derived from religious doctrines or "laws of nature," and there is no infallible way to resolve these arguments.

There are guidelines that may provide some guidance in choosing among competing behavioral standards. Constitutional law presumably sets a higher standard than conflicting statutory law, although the U.S. Constitution as interpreted for decades was used to invalidate child labor laws, and many would argue that the Constitution provided the lower standard. Another guideline is that statutory law presumably establishes a higher standard than a profession's code of conduct. Thus, society's interest in competition as a method of promoting consumer welfare, as embodied in the antitrust laws, seems generally to reflect a more compelling interest than a profession's interest in restricting competition through devices such as limitations on advertising, even when these devices are justified on the basis of the public interest.[15] Another guideline is that competing behavioral standards should be evaluated on the basis of logic: e.g., a standard used to evaluate a particular type of conduct should be internally consistent and should be consistent with behavioral standards used to evaluate similar types of conduct.

The resolution of conflicts among competing behavioral standards will vary by the type of conduct under consideration. Efforts to amend the Constitution will involve a different set of competing standards and a different set of guidelines to resolve conflicts than efforts to amend a city ordinance. The sets of behavioral standards and of conflict-resolving guidelines that pertain to particular types of conduct are not well-established, but I will offer some factors that seem relevant for determining the ethics of strikes by state and local government employees.

The U.S. Constitution provides little guidance for public sector strikes; they are neither protected nor prohibited activity. Likewise, federal legislation does not directly regulate state and local government labor relations. Public sector strikes are illegal in most states as a result of statutory enactment or court decision, although in eight states at least some public employees can strike if statutory preconditions are met.

What are the competing behavioral standards that can be used to evaluate public sector strikes? There is no professional organization that encompasses all state and local government employees that can be consulted for guidance, and organizations that represent segments of the public sector labor force divide on the strike issue, with unions generally supporting the right to strike and many employee associations discouraging the approach. Even if there were a unitary viewpoint among public employees on the strike issue, I believe that a jurisdiction's law should override a contrary view from a profession whose conduct is the subject of the dispute.

Some states have legalized public sector strikes, and other states deem such strikes illegal. Can logic be used to demonstrate the superiority of one of these stances and therefore establish an ethical judgment for public sector strikes that is applicable to all jurisdictions? Wellington and Winter used logic to demonstrate the differences between the public and private sectors and therefore justify a public sector strike ban. Krider and I used some empirical observations to challenge the logical propositions of Wellington and Winter, and we concluded that a differential strike policy for the public and private sectors was unwarranted. I obviously prefer one set of participants in this debate, but I do not feel the case for legalizing public sector strikes is based on unassailable logic.[16]

My review of the competing standards that can be used to evaluate public sector strikes leads me to the conclusion that for the purpose of ethics, the relevant standard is the jurisdiction's law. Public sector strikes that are legal are ethical, and vice versa. I believe that most public employees should have the right to strike, but the strike issue has been widely debated at the state level, and proponents of the legalized strike have had an opportunity to present their case. Where public employees and their supporters have failed to persuade legislatures to legalize the strike, acts that violate the ban on strikes are unethical.

The Desirability of Public Sector Strikes, Revisited. I conclude that the legal policy a jurisdiction adopts is the appropriate standard for deciding the ethics of strikes by the jurisdiction's public employees. In preceding sections, I argued that the desirable policy is the legalization of public sector strikes. However, I believe we are now at a point where the desirability of a ban on public sector strikes needs to be reconsidered.

I believe the desirability of the strike ban depends on the alternative offered to the Strike Model. If the alternative is a No-Strike Model in which the ultimate way to resolve disputes is unilateral decision making by management (subject to the restraints of the labor market), then I believe the Strike Model is preferable. However, the policy choices for the public sector ought not to be confined to the Strike Model and the No-Strike Model. An alternative that deserves consideration is the Third-Party Procedure Model, which refers unresolved disputes over working conditions to binding arbitration. There are advantages and disadvantages of the Third-Party Procedure Model, many surveyed in this essay.[17] I believe the evidence beginning to accumulate suggests that collective bargaining without strikes can involve meaningful negotiations and produce settlements that do not markedly differ from those resulting from collective bargaining when strikes are legal.

While the dispute-resolution approach which relies on third-party procedures has some disadvantages, such as the use of an outsider to decide matters about which the employer and union have more knowledge,[18] it should be stressed that the Strike Model of collective bargaining also has disadvantages. In addition to some of the problems already surveyed, another consequence of using the Strike Model is that unions do not uniformly gain strength when the strike is legalized. Most of the previous discussion implicitly contrasted the Strike Model and the No-Strike Model in terms of the consequences for the average union and the average employer. However, the usefulness of the strike to the union will vary considerably by the industry, occupation, and region in which the union operates. Some unions, as a practical matter, gain very little bargaining power from the right to strike. That legalizing the strike helps some workers more than others is a troublesome aspect of our policy in both the private and public sectors. But perhaps the result is more disturbing in the public sector, where I am uneasy about the notion that the economic and non-economic status of different groups of workers should be related to the degree of influence the groups derive from use of the strike.[19] The Third-Party Procedure Model at least takes us one small step beyond the idea that might makes right.[20]

VI. CONCLUSION

We may be witnessing the commencement of a new era in industrial relations—an era in which there is increasing disenchantment with the use of the strike and a willingness to explore other methods of dispute resolution.

If this era is upon us, it has been triggered by the explosion of unions in the public sector that has collided with the sector's traditional no-strike ban. To some extent, the essentially incompatible ideas of collective bargaining (which until a decade ago was almost automatically equated with the Strike Model of collective bargaining) and the no-strike policy for public employees have been accommodated by legalizing or tolerating public sector strikes. But to a greater extent, the incompatible ideas are being accommodated by developing a new collective bargaining model that relies on third-party procedures to resolve impasses. To our fascination and surprise, if not horror, the Third-Party Procedure Model of collective bargaining may be working in the public sector. And if it works there for enough years to convince skeptics that more is involved than the brief conjugal bliss that sometimes occurs in dictated marriages, pressures will inevitably arise to transplant the Third-Party Procedure Model to the private sector.[21] If that should occur, the result could be a fundamental restructuring of private sector labor relations that never would have transpired solely as a result of developments within the sector.

What a wonderful twist that would be. At the beginning of the 1970's, industrial relations practitioners and scholars were obsessed with deciding how much of the private sector bargaining experience could be transferred to the public sector. By the end of the 1970's, the new mania may be: How much of the public sector experience can be transferred to the private sector? *Posset cauda quassare canem?*[22]

NOTES

1. An exception to the textual statement involves those private sector employees with government contracts who must pay prevailing wages because of the Davis-Bacon Act.
2. Similar discussions of the advantages and disadvantages of collective bargaining are in Wellington and Winter (1969, pp. 1111–15) and Burton and Krider (1970, pp. 419–20).
3. An example of an economist who feels the advantageous non-economic

consequences outweigh their adverse economic effects is Rees (1962, pp. 194–95).

4. The virtues and vices of strikes in the public and private sectors are considered in the exchange between Wellington and Winter (1969) and Burton and Krider (1970).
5. A catalogue of the methods other than the strike by which public sector unions gain influence is included in Burton (1970, pp. 477–78).
6. The procedures for national emergency disputes under the Taft-Hartley Act provide one of the uses of fact finding in the private sector. The fact finding is unusual because the report on the dispute cannot include recommendations for settlement. Fact finding also is conducted by the emergency boards operating under Railway Labor Act procedures. According to Charles M. Rehmus, under the Railway Labor Act, "There have been 187 fact-findings, mostly since the 1930s, and they have largely been quite effective." (Correspondence from Rehmus, February 28, 1977.) Rehmus is co-author of a forthcoming study of the first fifty years of the Railway Labor Act.
7. According to a recent study by the U.S. Department of Labor, Bureau of Labor Statistics (1975, p. 4), grievance procedures are found in about nine out of ten agreements at the state and local government levels, a frequency "below the almost universal prevalence of such provisions found in private sector agreements, (but) measurably above the 82 percent rate found in Federal agreements."
8. Some arbitrators will abrogate contract terms that they believe violate constitutional or statutory law. See generally Meltzer (1967), Mittenthal (1968), Meltzer (1968), Howlett (1968), McKelvey (1971), and Meltzer (1976).
9. Evidence indicating that binding arbitration has virtually eliminated strikes by police and fire fighters is available for Pennsylvania (Stern, 1975, p. 32); Wisconsin (Stern, 1975, p. 112); New York (Kochan, 1976, p. 3); and Michigan (Stern, 1975, p. 71) and (Rehmus, 1974, p. 309). In correspondence of February 28, 1977, pertaining to Michigan, Charles Rehmus reported, "The 1976 data show that we still have no strikes relating to arbitration awards among our public safety officers."
10. See, e.g., the studies by Fogel and Lewin (1974) and Smith (1976).
11. A recent study by Ashenfelter (1976) provides some evidence that the union impact on relative wages in government is roughly comparable to the union wage impact in the private sector. The evidence is inconsistent with the conclusion in Lewin's survey article, which was prepared before the Ashenfelter study was available.
12. As Krider and I did in our earlier article, I am proposing to legalize public sector strikes only for local government employees not in essential functions. My research has focused on local government employees, and I do not want to generalize from that experience to state or federal employees.

13. According to Tsanoff (1955, p. 12), "Ethics is concerned with judgments of approval or disapproval, with the norms of valuation in human conduct."
14. For an interesting discussion of the "higher law," particularly as it pertains to American constitutional law, see Corwin (1955).
15. Ethical Consideration EC 2-9 of the American Bar Association's *Code of Professional Responsibility* asserts, "The traditional ban against advertising by lawyers, which is subject to limited exceptions, is rooted in the public interest. Competitive advertising would encourage extravagant, artful, self-laudatory brashness in seeking business and thus could mislead the laymen." By that test, all advertising should be banned.
16. Indeed, one of the points that Krider and I asserted is that logic is an inadequate guide for deciding the public sector strike issue, and that experience is also a relevant criterion. See Burton and Krider (1970, p. 418).
17. A possible consequence of the Third-Party Procedure Model is that unions will become more attractive to workers, since they will no longer be expected to participate in strikes. This may be a particularly important consideration for white-collar workers, many of whom consider concerted activity such as strikes inconsistent with their status. The increased attractiveness of unions that flows from the abolition of strikes would be considered an advantage of the Third-Party Procedure Model by those who support unions and collective bargaining.
18. A more thorough discussion of the disadvantages of statutes that use arbitration to resolve public sector bargaining disputes is provided by Clark (1974). Clark, who previously supported a total ban on public sector strikes, now supports legalization of strikes for most public sector employees if the alternative is a statute mandating arbitration.
19. Ted Clark has expressed a concern that arbitrators will give public employees with weak bargaining power higher wages than they could obtain from bargaining and will give employees with a strong bargaining position what they would have achieved if they had the right to strike. (Correspondence from Clark, March 1, 1977.) The result would be higher public employee wages under the Third-Party Procedure Model than under the Strike Model. I assume that arbitrators would boost the wages of workers with limited bargaining power and restrain the wages of workers with substantial bargaining power, compared to the wages that would result in the Strike Model. Clark and I both lack "solid statistical data" (to use his term) to support our positions. If Clark is correct, the case for the Third-Party Procedure Model is weakened.
20. Although, as Paul Gerhart has noted, the Third-Party Procedure Model does involve the use of "might" to determine working conditions in the sense that the arbitrator has power to dictate a settlement that would not occur in the No-Strike Model or the Strike Model. My argument implicitly assumes that the power exercised by arbitrators is more benign than the power unilaterally exercised by at least some employers

in the No-Strike Model or the power used by unions and employers in the Strike Model.

21. It is somewhat misleading to identify the experience to date with the Third-Party Procedure Model with the public sector, since the railroads and urban transportation have a long history of statutorily mandated third-party procedures. Nonetheless, if the Third-Party Procedure Model does become significant for the bulk of the private sector, the most likely impetus will be the experience in the public sector.
22. I am indebted to Jennette Rader, librarian at the A. G. Bush Library of the University of Chicago's Industrial Relations Center, for the translation of the profound inquiry, Can the tail wag the dog?, into dog Latin.

REFERENCES

American Bar Association. *Code of Professional Responsibility and Code of Judicial Conduct*. Chicago: ABA, April 1975.

Ashenfelter, Orley. "Union Relative Wage Effects: New Evidence and a Survey of Their Implications for Wage Inflation." Mimeographed. Princeton, N.J.: Industrial Relations Section, Princeton University, August 1976.

Berkowitz, Monroe. "Arbitration of Public Sector Interest Disputes: Economics, Politics and Equity." In *Proceedings of the 29th Annual Meeting, National Academy of Arbitrators*. Washington: BNA Books, 1976.

Boskin, Michael J. "Union and Relative Real Wages." *American Economic Review* 62 (June 1972): 466–72.

Bureau of Labor Statistics. *Grievances and Arbitration Procedures in State and Local Agreements*. U.S. Dept. of Labor Bulletin 1833, 1975.

Burton, John F., Jr. "Can Public Employees Be Given the Right to Strike?" *Labor Law Journal* 21 (April 1970): 472–78.

Burton, John F., Jr. "Local Government Bargaining and Management Structure." *Industrial Relations* 11 (May 1972): 123–40.

Burton, John F., Jr., and Krider, Charles E. "The Role and Consequences of Strikes in Public Employment." *Yale Law Journal* 79 (January 1970): 418–40.

Burton, John F., Jr., and Krider, Charles E. "The Incidence of Strikes in Public Employment." In *Labor in the Public and Nonprofit Sectors*, edited by Daniel Hamermesh. Princeton, N.J.: Princeton University Press, 1975.

Clark, R. Theodore, Jr. "Legislated Interest Arbitration—A Management Response." *Industrial Relations Research Association, Proceedings of the 27th Annual Winter Meeting* (1974).

Corwin, Edwin S. *The "Higher Law" Background of American Constitutional Law*. Ithaca, N.Y.: Cornell University Press, 1955.

Feuille, Peter. "Final Offer Arbitration and the Chilling Effect." *Industrial Relations* 14 (October 1975): 302–10.

Fogel, Walter, and Lewin, David. "Wage Determination in the Public Sector." *Industrial and Labor Relations Review* 27 (April 1974): 410–31.

Horton, Raymond D. "Arbitration, Arbitrators, and the Public Interest." *Industrial and Labor Relations Review* 28 (July 1975): 497–507.

Howlett, Robert G. "The Role of Law in Arbitration—A Reprise." In *Proceedings of the 21st Annual Meeting, National Academy of Arbitrators*, edited by Charles M. Rehmus. Washington: BNA Books, 1968.

Kochan, Thomas; Ehrenberg, Ronald; Baderschneider, Jean; Jick, Todd; and Mironi, Mordehai. "An Evaluation of Impasse Procedures for Police and Fire-fighters in N.Y. State: Survey and Recommendations." *N.Y. State Public Employment Board* 9 (November 1976): 2.

Lewin, David. "Public Sector Collective Bargaining and the Right to Strike." In *Public Employee Unions: A Study of the Crisis in Public Sector Labor Relations*. San Francisco: Institute for Contemporary Studies, 1976.

Lewin, David. "Public Sector Labor Relations: A Review Essay." *Labor History* 18 (Winter 1977): 133–44.

Lewis, H. Gregg. *Unionism and Relative Wages in the United States*. Chicago: University of Chicago Press, 1963.

McKelvey, Jean T. "The Presidential Address: Sex and the Single Arbitrator." In *Arbitration and the Public Interest, Proceedings of the 24th Annual Meeting, National Academy of Arbitrators*, edited by Gerald G. Somers and Barbara D. Dennis. Washington: BNA Books, 1971.

Meltzer, Bernard D. "Ruminations about Ideology, Law and Labor Arbitration." In *The Arbitrator, the NLRB, and the Courts, Proceedings of the 20th Annual Meeting, National Academy of Arbitrators*, edited by Dallas L. Jones, pp. 1–20. Washington: BNA Books, 1967.

Meltzer, Bernard D. "The Role of Law in Arbitration—A Rejoinder." In *Developments in American and Foreign Arbitration, Proceedings of the 21st Annual Meeting, National Academy of Arbitrators*, edited by Charles M. Rehmus. Washington: BNA Books, 1968.

Meltzer, Bernard D. "Arbitration and Discrimination: The Parties' Process and the Public's Purposes." In *Arbitration—1976, Proceedings of the 29th Annual Meeting, National Academy of Arbitrators*, edited by Barbara D. Dennis and Gerald G. Somers. Washington: BNA Books, 1976.

Mittenthal, Richard. "The Role of Law in Arbitration." In *Developments in American and Foreign Arbitration, Proceedings of the 21st Annual Meeting, National Academy of Arbitrators*, edited by Charles M. Rehmus. Washington: BNA Books, 1968.

Rees, Albert. *The Economics of Trade Unions*. Chicago: University of Chicago Press, 1962.

Rees, Albert. "The Effects of Unions on Resource Allocation." *The Journal of Law and Economics* 6 (October 1963): 69–78.

Rehmus, Charles M. "Legislated Interest Arbitration." *Industrial Relations Research Association, Proceedings of the 27th Annual Winter Meeting* (1974): 307–23.

Ryscavage, Paul M. "Measuring Union-Nonunion Earnings Differences." *Monthly Labor Review* 97 (December 1974): 3–9.

Scheible, Paul L. "Change in Employee Compensation, 1966 to 1972." *Monthly Labor Review* 98 (March 1975): 10–16.

Slichter, Sumner H.; Healy, James J.; and Livernash, Robert E. *The Impact of Collective Bargaining on Management.* Washington, D.C.: The Brookings Institution, 1960.

Smith, Sharon P. "Government Wage Differentials by Sex." *Journal of Human Resources* (Spring 1976): 185–99.

Stern, James; Rehmus, Charles; Loewenberg, Joseph; Kasper, Hirschel; and Dennis, Barbara. *Final Offer Arbitration.* Lexington, Mass.: D. C. Heath, 1975.

Tsanoff, Radoslav. *Ethics.* Rev. ed. New York: Harper & Brothers, 1955.

Wellington, Harry H., and Winter, Ralph K., Jr. "The Limits of Collective Bargaining in Public Employment." *The Yale Law Journal* 78 (June 1969): 1107–27.

9

PUBLIC SERVICE STRIKES:

Where Prevention Is Worse Than the Cure*

VICTOR GOTBAUM

Some years ago, I came across an old friend who became a professor at the University of Wisconsin. He had good practical trade union experience which he converted into bad professorial habits. This became painfully apparent to me when we sat down and he said to me, "Vic, I want to test a new idea," and he went into third-party procedures. It was a magnificent Rube Goldberg. It mixed mediation, fact finding, arbitration, and voluntary arbitration. He went through all this tortured stuff and, in order to give the workers equity (in all fairness, his sympathies were with the workers), it became more and more tortured in terms of making sure the mechanism was impartial. I asked, "Why are we going through this?" "Well, it's an alternative to a strike." I looked at him and said, "You never had this unhappy kind of reasoning before." He looked at me nervously and let me know that now he had to look at it academically, and even more importantly, he said, "Vic, we have to face the fact that strikes are illegal in the public sector and we've got to be aware of this." He went through another seven minutes of that. I was pretty good about it. I said, "Knock it off. What you're doing basically is cementing your prejudice. The truth is, third-party procedures are killing collective bargaining in the public sector. They have become tortured. Although I would agree with you that there is a possibility of change, at this time they are so one-sided that there are just a few areas where these procedures are impartial." Some people

* This paper was presented at a symposium held at the University of Kansas in November, 1976. Questions from symposium participants and responses by Mr. Gotbaum are included in the text.

want to set up a mechanism to make certain that public employees do not do something illegal; they want to keep workers, you might say, ethical. And I would say this is nonsense.

What has really brought it about? Let's look at the atmosphere, the environment; that's what has brought it about. We are really catering to one of the worst prejudices of the American public—their disdain for the public sector, their artificial belief that the government that governs least governs best. They do not like public employment. They do not like public service. They do not like public employees, and God forbid, you should unionize them. That would be even worse. They look at all sorts of methods and ways of stultifying collective bargaining and keeping us away from the traditions of the private sector. What makes it even sadder is that the public sector is growing and it is kind of like "out, damned spot"; but the damned spot keeps growing.

One out of every six and a half workers is now a public employee. By the twenty-first century, we expect one of four will be a public employee, despite depression or recession. This makes it worse, because the public needs us and they want us. They want smaller classrooms and they have to hire more teachers. If the public wants their middle-class cars to have a smooth road, they need more public laborers. So they need us, but they do not like us. The more they need us, the less they like us, and as a result, it is very simple to dislike the public employee.

Well, what has it created? The way I see it, and what I find so distressing in terms of ethics, is that nice guys wonder how they can prevent strikes. Rather than looking at good, constructive collective bargaining and how we can professionalize it, we set up procedures that stultify these amateurs who just entered the union market, both labor and management. We prevent them from becoming professionals. That is one aspect.

The other aspect of it that really worries me is that strikes have become only the fault of the *workers*. Deficient, amateur management, an unproductive mayor, stupidity on top are never blamed for the strike. It is either the teachers who go out because of some very deep frustrations, or the social service workers or laborers who are told that they are always at fault. There is really no careful examination under the Taylor Act, for example; there is no way of punishing management. There is a clause in the Taylor Act in terms of culpability according to which fines or punishments against striking employees can be reduced if management is found to be guilty of extreme provocation. But we never can get the official who caused it in front of a court. The "extreme provocation" argument has never really been brought into play. It is always the unions, the workers, who are at fault. And this, I say, also creates the atmosphere.

There is no inherent difference between the public and private sectors. Let me make just a few points to emphasize this.

I have a hunch that people who are antilabor in the public sector and want to kill the strike there would also like to kill it in the private sector. It is just a hunch. It has not been whipped up yet, but we are coming close.

In the teachers' strike and sanitation strike in New York City in 1968, for example, both leaders, Al Shanker and John DeLury, were jailed. Both unions were fined. The teachers lost the dues check-off, and that literally cost them millions of dollars. Neither strike had a terrible effect on the economy. In fact, lots of people felt that the kids needed a rest from the teachers, the way the school system was going. John DeLury struck, I think, in the middle of winter, and if you went through Harlem, nobody noticed the difference with the strike or without it. So the strike had little effect. Nevertheless, the leaders were jailed, the unions fined, and so on. In the same period, a little after that, Con Edison went out on strike, which meant there could have been death and everything else if the blackout occurred. The fuel drivers went out on strike, and there were deaths directly attributed to it. The elevator operators went out on strike, and with the high rises in New York a lot of the old people got caught on top. Nobody got upset by it. Not a word. It becomes ridiculous.

Just recently, Local 1199, which represents the *private* hospitals in New York, went out on strike and our union went out on strike. Ours was shorter. Everyone admitted that our strike was justified, that the union acquitted itself nobly, that we saved jobs by making sacrifices. Those who knew the situation (and I guess I'm a little prejudiced on it) admitted that it was stupid management that forced the workers out. Nevertheless, the workers had lost a week's pay, and we are still in court. But in the 1199 strike, nothing was done about it. It involved basically the same workers, serving the same clientele, but nothing was done about it.

So what happens is that despite all this talk, we in the public sector have been "impassed" to death. I'm not against honest third-party procedures. But the truth is that the third-party procedures have been invoked to deny workers in the public sector militancy. When this is done, almost by definition they become counterproductive and managerial in content. The Taylor Act is not an exception to the rule. This act has all sorts of mechanisms in it. But, in the final analysis, in this impasse procedure the boss decides whether the boss is wrong or right.

It would be like General Motors asking Ford to make a final decision in the dispute between them and the United Auto Workers. It is that crude. In the Taylor Act, the legislative body has the final say. Everybody knows that the legislative body will, in 99 percent of the cases, rubber-stamp the administrative decision of the boss. But in other areas, it is done with more sophistication. That is, the boss selects the impartial arbitrators. The boss cannot miss; he gives the guy $35,000 and selects him. If the arbitrator makes a few wrong decisions, he's out. That is basically the kind of impartial stuff they've set up. We have been hurt by this, terribly restricted, and what I am saying is that it has killed collective bargaining.

There is little collective bargaining in the public sector. There is power and politics, but almost no collective bargaining. In many cases, a strike itself might have little effect. But the right to strike does. This insistence, this obsession with preventing strikes, hurts the unionized public employee more than anything else.

And I would beg of you not to make wage comparisons to non-unionized workers to decide whether the strike is more effective or collective bargaining is more effective. Wages as an absolute measurement is terribly materialistic, almost too materialistic. That undefined substance known as "dignity of work"—the man or woman on the job being able to stand up against the boss on an equal basis—that, to me, is much more important. That is where I think our unions have succeeded far more than in the absolute wages.

We need the possibility of the strike. Take what I call the Red China experience. When the Red Chinese did not have the atom bomb, they were the most militant and irresponsible of sovereign states. They wanted Russia to invade the United States; they were flexing their non-existent muscles. But as soon as they got the bomb, the Chinese became much more responsible, because they could also create havoc. The same thing happens with workers. The most militant workers with whom I have to contend are the social service workers. They are the greatest militants in the world. They are going out on strike every Monday and Thursday, and you know why? Because nobody takes them seriously. So they are always striking. My good friend Lillian Roberts calls them "mouth militants." They are the most magnificent of mouth militants. They do not have the atom bomb. But our sewer laborers, who could pollute the waters of New York City in three days, almost never talk strike, because they know the power that they have.

Those who look for examples of the havoc created by strikes in the public sector literally have to go back to the Boston Police Strike, which made Calvin Coolidge a great man. It is very difficult to find another strike that really created this kind of havoc in the public sector. Workers are very, very careful in that regard. They know what their strength is and where their strength is. They are very, very cautious indeed. By the way, the same amount of caution exists in what I call marginal workers or workers who know that if they pull the chain, it can cost them X number of jobs. I do not talk strike in New York City anymore. Not because I have become less militant, but because I want to represent in a constructive fashion the men and women who pay me, and I do not want to kid them about strikes, etc. I know if I pull the chain now, it can mean default and 50,000 jobs, or closing down thirty more manufacturers, who will move to either Oshkosh or New Jersey or, worse than that, the Deep South.

Big Steel. In all fairness to I. W. Abel, I do not think, despite my love for Ed Sadlowski, that Abel became less militant and toned down on the right to strike. It hurts me to say this because I happen to like Ed Sadlowski. The sad truth is that in the last strikes, the increase in imports and steel from other areas was so overwhelming that it made I. W. Abel very, very nervous indeed. If Sadlowski had won the Steelworkers' election, I have a hunch he would have toned down his militancy, possibly after one strike. I am mindful of a magnificent statement by Péguy, the French philosopher, that a conservative is a socialist who becomes an administrator.

So, as a practical trade unionist, I'm not concerned about the ethics of strikes. The burden of my remarks is not to prove that the unions represent only virtue. Not at all. We represent vice, we represent virtue, we represent workers whom we believe deserve a living wage and deserve dignity on the job. We use measures that sometimes might upset many people. I do not deny it. But unions, collective bargaining, are an exceptionally fine American tradition—not only for workers, but for the American economy as a whole. It has been a successful system. Our standard of living has grown. We have a decent economy. Despite strikes and animosities and confrontations, we somehow manage to live together, get along well, and do the job.

There is no difference at all, really, between the public and private sector. I want the strike weapon to remain intact. There will be less use of it if we feel it's not going to help the workers, but I want the right to use it. There has been no overwhelming suffering because of public strikes, and I would hope that we do not institute laws that begin to sovietize us. For, let me remind you, strikes are illegal in the Soviet Union.

Question: If we have to pick one of the outstanding examples of the use of third-party procedures, it is the steel industry, where the unions and management have voluntarily agreed that, for a three-contract cycle (nine years), they waive the right to strike and agree that in case disputes are still outstanding at the end of the collective bargaining period, they will be referred to binding arbitration. Now, I realize that's one of the issues in the campaign; but the point is, I do not see that there is any killing of collective bargaining in the steel industry because they have gone to third-party procedures.

Gotbaum: In steel, I have no quarrel with I. W. Abel; but he waived his right. What a man gives, he can take back. It is not the same as the law saying, "You do not have the right. It is illegal. We are going to kill your firstborn." It is quite different. I. W. Abel knows he can always call on it, and the struggle within the Steel Workers is very healthy. I think the struggle is good, but quite different because in those areas where you are forced to give it up, collective bargaining is killed.

Question: Regarding the future alternatives for the public sector, your Red Chinese example is well taken. It has been my experience that most union leaders today are reasonable. They realize their power, and they work very closely with management. It has also been my experience that they have a difficult time with some of the people they represent. I have had to swallow three bitter pills as an executive of a company with a dozen or more businesses where we were made uncompetitive because of a strike. We knew it was going to happen. We tried to negotiate, but we finally made a settlement that just put us in a position where we finally had to close down. In the private sector, you can close down. You can button up a plant, pay off, take your money and put it into something else, but in the public sector the public cannot strike against the employees—the public employees, that is. The public just has to grin and bear it. In the future, will the government reduce the number of employees that they have and subcontract these various services, or will they be limited by strike under state law, or will they continue to be run by labor leaders who are reasonable? How is it going to go?

Gotbaum: I do not know how it is going to go. I care, of course. But we are dealing with human beings, and I do not think that we have a right to insist on some kind of monolithic pattern. I have been insisting on (and I've heard this from Jerry Wurf, our national's president): the union is a union is a union, and do not expect workers to be different. Do not expect the leaders to be different. Do not expect us to be "reasonable" and "nice guys." I think sometimes people demand virtues that public sector leadership cannot have and ought not to have. Construction companies cannot move out. Most of the industries cannot move out. If we strike, we hurt the citizen; what strike does not hurt the citizen?

I get captivated by some of the erroneous comparisons people make. An automobile can increase in price 300 percent. Your food can go up 200 percent. If your taxes go up even less of a percentage, somehow the public is being raped by public employees. That is not so. In fact, our own studies show that the wage bill has not been going up that high since the arrival of unionism, taxes have not increased at a greater pace than costs in other areas, and yet we get this funny comparison that somehow when workers in the public sector strike, they get a helpless, hopeless citizen. If the Teamsters strike, or the Building Trades workers strike, or the Auto Workers strike, or the Garment Workers strike, they are not hurting a helpless public. The industry can take it. You know as well as I that 90 percent of industrial movement away from a geographical area has little to do with strikes. The movement because of a strike is so minuscule, so small, that you can't even find the percent for identification. So that argument does not hold water.

I hope public employee labor leaders, as well as private sector labor leaders, become responsible. I hope we do not collude, we do not discriminate; I hope we are sensitive to the public. If we fail, we should be punished for it.

Question: But would private industry take over on a subcontract basis some of the services and diminish the number of public employees?

Gotbaum: You know why public service comes about? Because the private sector fouls it up in the first place, or they do not want it. We forget that there's a peculiar belief that the private sector could come in and do the job better. We have a growing public sector either in areas that the private sector has fouled up or that they do not want. Now we get a sort of *deus ex machina* where the private sector's going to come in and do the job better. The reason the public sector is growing, quite candidly, is because the private sector has not been doing what it could. Why did the state university grow? I hear the most nonsensical arguments about how the voluntary hospitals ought to take over the public hospitals in New York. The reason the public hospitals came about in the first place is that way back, the sick poor were dying in the streets because the voluntary hospitals would not take them in. Now they want these same voluntary hospitals to do it because some kind of—we're talking about ethics—new ethical conduct is taking hold of the doctors and the surgeons and the hospital administrators. Nonsense. They still have disdain for the poor, they still segregate them within their hospitals, they still set up a two-class system. So I am really not worried about contracting out. I am worried about the myth and the fact that we may set up white elephants. We have seen it in some areas. They'll bid low, take it over, and then, when the public aspect of it is out, watch the price. You'll forgive me, but I do not trust them.

Question: Do you feel that police and fire forces should have the right to strike? Do you see any limitations on the right to strike on the part of any public sector employees?

Gotbaum: Our union [The American Federation of State, County and Municipal Employees, AFL–CIO] stand is that there should be arbitration for police and fire. It's done on a very pragmatic basis. We won a National Labor Relations Act, and Jerry Wurf justifiably says, "Vic, if we come out for an act in which police and fire are given the right to strike, then we're kidding ourselves." I do not argue with him. It is, I think, for the greater good, and since police and fire themselves in the main do not really want it and have asked for arbitration, I do not fight it. But it is a contradiction, in my own belief. Truthfully, I believe everyone should have the right to strike. But if you withhold your labor, you withhold your labor very carefully and gingerly indeed. It is no accident that our union, for example, has the smallest per capita rate of strikes of any union in New York City. We're talking of a right. Pragmatically, by the way, and nationally we've stated: in services that can kill you, substitute arbitration. You notice, I still cannot say it, I cannot say, "Take away the right to strike." Substitute arbitration. I keep saying that, and we do it on a very pragmatic basis.

You get some funny reactions, I have to tell you, in terms of the workers themselves. Here is a classic example that I will never forget. In the height of the summer—and this is almost a life-and-death situation in New York, because you know the need for the public beaches in Coney Island—our lifesavers had had it. They felt they were put upon. They did not have enough respirators. They were really quite upset, and they let me know they were going to strike. I said, "You can't do this. The people are going to use the beaches. You can't have kids in the damn ghettos. It is really hot, and you just can't do this, fellows." They said, "Vic, don't worry. We'll strike, but we'll stay on guard." So they got off their perches and talked to the girls, which I suspect they did also when they were not striking, and they were waiting for someone to yell. It was sort of interesting. These lifeguards could not bring themselves to walk off the beaches. So it was basically a phony strike.

I bring this up because workers are very sensitive in terms of what they can and cannot do. I think in Kansas they had a fire fighters' strike where the fire fighters stayed near the fire stations. They did not leave the fire stations unguarded. It is sort of fascinating. In our hospital strike, by the way, the workers were told to let the doctors through so that if there was a real emergency, they could take care of it.

Question: What about the New York police, in the recent turmoil, who are said to have encouraged the violence?

Gotbaum: I think you are talking about the trouble around Yankee Stadium during the Ali-Norton fight. Did they act irresponsibly? Yes, we thought they acted irresponsibly. I stated to you that we are not always virtuous, and I am not going to argue the point. I think that we behaved rather badly. But the frustrations of the average policeman are immense. The cop is given an impossible job. He is told that he can prevent crime, and he knows damn well that he cannot.

Question: How do you determine when to go on strike? You use the expression, "If I decide to pull the chain." Do you really mean that you have that kind of power, or do you mean that you get an expression of workers' desire for a strike before you pull the chain?

Gotbaum: You are right, and I should be criticized for using the first person singular. We represent frustrations—for example, the hospital strike in which we were involved just recently. We had to go out on strike. The frustration of the workers was unbelievable. In fact, that was one I did not want. Take the last teachers' strike of Al Shanker of the UFT, United Federation of Teachers. He tried to plug that one up. You never saw a man work so hard trying to plug it up. He could not do it. He really could not do it. So, you'd better pay attention. When the frustrations grow, you have to pay attention.

Question: So you're saying you do have a strike vote?

Gotbaum: Yes.

Question: Mr. Gotbaum, you are engaged or have been involved or associated with strikes that have been said to be illegal. Obviously, you've had some considerations within your own mind and own ethical system as to whether or not it's appropriate to engage in such strikes. What kinds of considerations do you make in determining whether or not to engage in an action which, with respect to a statute, is outside the bounds?

Gotbaum: The first consideration is, can I win? That is the first: can I win the strike? And what brings me to that strike? I have to look at what it means to working men and women. This is my advice, in a sense. All of us call what we do "for the public good." That's nonsense sometimes. On the other hand, I have to look at the workers whom I represent and see where we are going.

Let me tell you about once when I was too ethical. I made a mistake for which I will never forgive myself. We organized the workers in the Chicago hospitals and did a magnificent organizing job. But we could not get recognition. The conditions were incredible, and I will not burden you with them. So, although we had in one hospital 85 percent of the people signed up and in another 90 percent, they would give us no recognition. We went out on strike. Now, I could have won that strike if I had not listened to my lawyer, a magnificent human being, Abner Mikva. He was very concerned that if we did something really wrong, some of the people would have been jailed. On the other side was my present associate director, whom I call one of the country's natural great trade unionists, Lillian Roberts, who said a very simple thing: "Victor, we're at war. Our people have suffered. We've got to win this war regardless." She wanted to use violence against the scabs. I was torn. Ab prevailed. Because he did, we lost the strike. I will never forgive myself. I should have followed Lillian's advice. She was absolutely right. The strike lasted for five months; the workers went back under amnesty conditions. Some nine or ten years later, the Teamsters organized a union there. At any rate, Lillian's point was well taken. You fight, you scratch, you kick the hell out of some of those scabs, but you win it. Abner was wrong. If we had this much suffering and it was worth it, you roll the dice for it. Now, if you want to talk about ethics, some of the ideas that Lillian had were pure, illegal acts, but she could not have been more right. To this day, I look back at one of the mistakes I've made as a labor leader—it was following Ab Mikva's advice and being "ethical." I do not burden myself with that anymore. And I am no less or more of a son of a bitch than I was then.

Question: You said your colleague Lillian was absolutely right in the Chicago situation. Do I interpret that to be morally right or strategically right from the labor point of view?

Gotbaum: Well, obviously, we justified it. So I would say it was morally right. Here were workers making 65 cents to 85 cents an hour. Here were workers without dignity and recognition. Here were workers who were suffering a disease far greater than the people they tended in the hospitals. At least the people in the hospitals were there temporarily, and we would hope they would be extricated. I absolutely feel there was moral justification to it. They needed a union, they deserved a union, and they should have had a union. That's why I am still to this day upset with the decision I made. I do not think I was morally correct in worrying about the niceties of the law. I was in real pain when we put that strike to bed. I think I was using middle-class values that just did not apply to the situation. I am happy to say that I learned well from Lillian, and I will not make that mistake again.

Question: What's wrong with the middle-class values?

Gotbaum: In this particular case, they were consigning black and chicano workers to 65 cents an hour. They had no identity; they had no dignity on the job. They were treated like driftwood. What was wrong with middle-class values? It was a stench in the nostrils of decency.

Question: What about the beating up of scabs? How do you justify beating up the scabs, and how far do you beat them?

Gotbaum: So that they do not come back to work. You kick the hell out of them so they cannot get to work, and you make sure that in the long run, other people will work with dignity. That was the bad mistake I made.

Question: How far can you beat them, in your opinion, morally? Jimmy Hoffa, speaking on a similar question a few years ago, described a strike in Canada where the picket line was being broken because some priests were coming by and encouraging workers to go into the strike. He said, "We took care of those priests. We broke their knees with baseball bats." I think the problem we get into here is that we've got competing morals. It's always possible to find some moral justification for what you do, no matter what it is. If you got the director of the hospital in here, he could build a case that would be, to a lot of people, a morally persuasive case as to why the union had to be kept out.

Gotbaum: My country taught me how to kill. I was at first a little disturbed by it until I entered the town of St. Lo and saw what the SS did. I was delighted then that they taught me how to kill, and I became a terribly efficient soldier. I murdered, you know, cold turkey. I murdered and felt morally justified. I guess if you spoke to the victims' mothers and fathers, they'd be terribly upset; but this is really what it boils down to. I am not going to quarrel with you in terms of how the hospital management felt about it. Obviously, after we signed up 90 percent of the workers and they wouldn't give us recognition, they knew what they were doing, and then each guy looks into his own belly and decides where he goes on it. It does not mean that by definition I am correct; but I will state to you that I know that I am correct when I leave a situation and feel good about it. When I left that situation, and the workers had to go back (although we increased their wages by some 40 to 50 percent—if you want to use a wage comparison as to whether we were successful or not, we were evidently successful), I felt I did an immoral thing by not listening to Lillian. That's why I felt guilty. If I had broken some legs, I would have felt much better about it today.

I have to look at society as a whole and the contributions we make. I do not deny that—it's very hard. I do not care whether it's labor relations, marital relations, international relations—it's hard. It is hard, for example, in a war to find someone who says, "I'm the aggressor, and I'm damn proud of it." You never find this. In a broken-up marriage, you never find either party saying, "I'm responsible for the breakup of the marriage, and I've been a horse's ass." Where do you find it? In labor relations, it is even more pointed. In labor relations, where management and labor confront each other, did you ever hear GM say, yes, we were wrong, we should not have done this to the auto workers. So we move. This does not mean, once again, that we're virtuous. Earlier, in a moment of terrible objectivity I admitted that we do not only represent virtue, we represent vice as well. I do not deny it. Where are we going, what does it mean, what is the direction? These are the things we have to look at.

Question: Ethics, like pornography, might be a matter of geography. If there is any merit in that, what experience could we gain on this ethical consideration by looking to other countries—Italy, Germany, Canada—

that might have resolved the dilemma we have been discussing of public sector third-party procedures versus collective bargaining and the right to strike?

Gotbaum: No. Canada's laws are far less restrictive than ours. They give you a choice. They do not kill the first-born if the workers go out on strike. It is fascinating. In most countries, there just is not the attitude that we have here in the United States. In England recently, the dustmen, the sanitation men, and the railroads were shut down by a strike. Nobody burped. You did not have editorials in the *London Times* such as you would have in the *New York Times.* In Israel, there is a strike every Monday and Thursday. There are third-party procedures in West Germany, and my good friend, Heinz Kluncker, who heads up the Public Service union, probably the largest one in the world, had close relations with the Socialist government of Willy Brandt. He took Willy to the mat, strike after strike. He made me very proud of him.

I am not against third-party procedures. I am not arguing against the effective use of third-party procedures. I am arguing against their mandatory use with the negation of the strike. This disturbs me. And there is no country, interestingly enough, outside of the totalitarian countries—Soviet Union, underdeveloped areas—where the right to strike is taken away.

Question: Do you think there is merit in an ombudsman for the public, to represent the public in a strike situation?

Gotbaum: I do not believe you can mediate, arbitrate in public. The truth is, it is a much more closed operation than that in the give-and-take of negotiations, when the negotiating committee pushes and the other side pushes. The ombudsman approach just does not hold up.

Question: Does not the strike mechanism boil down to a question of might versus right? In terms of the strike, the side on which might exists is the one that's most likely to succeed. At least with arbitration, there is a better chance that the decision will go in terms of what is right rather than simply where the balance of power lies.

Gotbaum: Admittedly, there are imperfections. I do not deny this. But you get some sobering up. Look at what is happening in the building trades right now. They have been on strike for a month of Sundays, and you know why. These are marginal industries. These things even out. But look at Turkey, look at the Soviet Union, where workers do not have the right to strike. The choice of systems is clear.

Question: About the scabs whose legs were mercifully but perhaps, in your judgment, incorrectly not broken. What about their rights?

Gotbaum: Damn them.

Question: You come out strongly against mandatory third-party interventions, in the absence of which the interplay of force, largely peaceful market forces in most cases, determines the outcome of the strike. In the case that you cited, there were market forces at work. The scabs wanted to work, were willing to cross the picket line. If the market worked impersonally and peacefully, as it does in most cases and in most strikes, they would have gone to work. And yet, you feel it was a mistake not to invoke a different kind of power, physical power, to prevent them from operating in what they saw as their best interests. How do you justify that?

Gotbaum: Very simply. They were depriving people fighting for a living wage, fighting for the right to clothe their kids, pay their rent, to be able to say, "I am a man, I am a woman." They were depriving them of this right. So my considered judgment is that a few broken legs is far less vicious than the destruction of human beings. This is the way I saw it. But you have to remember that most evil is done in the name of decency. There is no question about it. "Don't interfere with the Soviet or the Nazi concentration camps because you are interfering with the sovereign rights of a nation. Let's keep out of it. Because after all, the sovereign rights of a nation within international law should be respected." Apartheid in South Africa. Every evil can be justified by moral standards or by a law to the person who perpetrates it. Then we make other judgments. Unions have struck for recognition. Otherwise, there would be no unions. The law has clearly militated against it. These are judgments we have to make. I was told it was legal for me to kill. It was legal for me to kill, to murder, under a nice, new, different set of circumstances and, by the way, I thought it was morally justified in terms of being part of the troops that invaded Nazi Germany. I had no problem with that one. I have no problem with breaking a few legs to bring dignity and unionism to tens of thousands of workers who need it.

Question: But do not the scabs have both dignity and independence?

Gotbaum: Do they? Do you want a reply from me? Is that what you really want, a reply from me? They have no dignity. They ought to be locked out. Okay? I feel this in the bottom of my belly. Do you remember Jack London's definition of a scab? They are so low they can crawl under a worm's belly. The man was poetic and beautiful. So when I am asked these middle-class questions about whether the "scabs are miserable"—they are.

V

ETHICAL LIMITS TO ADVERTISING

10

THE ETHICS OF ADVERTISING

BURTON M. LEISER

INTRODUCTION

Advertising is far more ancient, and in some respects more respectable, than many people imagine. In a sense, the first person to cut open a fruit or vegetable so that passersby could see its interior before buying it was engaged in advertising; and the peddlers who hawked their wares through ancient streets were croaking out ads of a kind, while those who invented tunes or chanted "Cockles and mussels, alive alive-o" were performing the first singing commercials. They performed a valuable service, for they enabled potential buyers to become aware of the availability of products or services which they might be interested in buying, the same function performed by our more sophisticated advertisements and commercials.

Ads bring buyers and sellers together. If it is not wrong to buy a product, then it can hardly be claimed that it is wrong to sell it. The merchant's function in society is vital. Even Plato accorded him a place in his ideal state, for it was obvious that under the most primitive and Spartan conditions, division of labor was essential. If everyone was to do his own job well, people had to be relieved of the irksome tasks of producing everything for themselves and of grubbing about aimlessly for the necessities of life. The merchants who brought their wares to early market places relieved people of the need to wander about, finding each of the many items they needed for themselves. The invention of a central location for shopping was probably as important to the growth of civilization as was the invention of the wheel, though it seems to have been overlooked by historians. And the success of such operations must have depended from earliest times upon some form of advertisement—that is, some means of attracting shoppers to the merchant's place of business, of "turning him toward" the merchant, as the

Latin origin of the word indicates. In this respect, advertising has changed very little.

At a very early stage of its development, the heart of merchandising came under various forms of governmental regulation. Such regulation can be explained by only one hypothesis: that merchants occasionally cheated their customers and dealt unfairly with one another. For no society passes laws to control practices that no one engages in. Thus, as early as Hammurabi and the Bible, we find regulations dealing with honest weights and measures; and early Jewish law interpreted the rule of *hassagat gevul* (literally, moving the boundary marker) that is found in the Bible as referring also to unfair encroachment on a fellow tradesman's territory.[1]

One of the important functions filled by advertisement is its stimulation of competition and the consequent reduction of prices. This might seem in some respects to be antithetical to the principle that underlies the biblical, talmudic, and later prohibitions against unfair competition. But the emphasis must be placed upon the term "unfair." Early legislators recognized that there was a difference between competition and unfair competition. Competition that increased the customer's options was to be encouraged, but not if it meant cutthroat tactics or the destruction of a man's only means of livelihood.

One of America's greatest admen, Bruce Barton, once described an ad he had written for an insurance company. Because of that ad, a young man from New Jersey purchased a life insurance policy that would guarantee his wife and three children an annual income. A few days later, he died. Barton wrote, "Many times in the intervening years I have been reminded that somewhere in New Jersey there are a mother and three children, now grown up, who, without the slightest suspicion of my existence, have had their whole lives changed by the fact that one day I put together some words that were printed in a magazine, and read . . . by their husband and father, who was influenced to do what I suggested."[2]

It would be unfair to suggest that advertising is, by its nature, either useless or bad. It is not. On the contrary, it has an important, useful function to fulfill in a free society. Anyone who has used the classified columns of newspapers to sell unwanted objects, to retrieve a lost dog, to seek a better-paying job, to find a handyman to do carpentry work that he was unable to do himself, to find fruits and vegetables less expensive than those he might purchase at the supermarket, or for any of a multitude of other purposes, would concede that there *are* ethical advertisers, that advertisements can work for the public good, and that society as a whole would be better off if certain restrictions on advertising were lifted. Truthful advertising can help bring people with common interests together so that they can achieve ends that they could not reach without one another. It is an important instrument that can encourage free trade and competition and be a powerful force for good in society.

Like most social practices, advertising cannot be assessed morally as if it were a single institution. Individual practices must be examined to determine whether they measure up to relevant ethical principles. These principles are basically identical to those that would apply to any other area of human conduct, and their justifications are essentially the same: A practice that contributes to human happiness and well-being, and reduces the amount of unnecessary suffering in the world, may be regarded as morally proper; while one that tends, on the whole, to reduce the measure of well-being in the world and to increase needless pain and suffering may—all things being equal—be morally condemned. Long experience has revealed that while certain kinds of actions may provide short-term advantages, they tend in the long run to have gravely adverse consequences for the population as a whole. Fraud and deception, for example, may be advantageous for a time—particularly for those who are concealing the truth—but they tend, in the long run, to diminish respect for the truth, to increase popular cynicism, to make communication more difficult, and to cause a breakdown in other important institutions. Moreover, such practices are seldom advantageous, even over the short run, to those who are being deceived.

It will be convenient to divide the subject into four major areas of concern, each of which raises distinct moral questions which will be considered in turn, together with appropriate examples from recent cases:

1. The nature of the product being advertised.
2. The person to whom the ad is addressed.
3. The impact upon persons who are likely to be indirectly affected by the advertisement.
4. The substance of the advertisement itself.

1. A SALESMAN'S RESPONSIBILITY FOR HIS PRODUCT

Any person who urges another to purchase and use a product or service assumes a responsibility toward him. The advertiser is not merely an innocent middleman who conveys a message from one person to another. He helps to create the message, using all the specialized skills of his art to persuade the potential consumer to act favorably upon his appeal. He shares in the rewards of successful advertising campaigns. He therefore assumes a responsibility for the product he induces the customer to purchase. In particular, if the product is dangerous or harmful, the advertiser who has persuaded the consumer to use it shares responsibility with the manufacturer for any harm that may result. This responsibility ought to be enforced by law, both with penal sanctions where the harm is particularly great and with appropriate remedies in tort. It is, in any case, a moral responsibility, for were it not for the advertiser's intervention, the consumer might never have suffered the damage done to him by the product he purchased.

By the same token, the advertiser has a right to feel, as Bruce Barton did, that he has contributed to the well-being of those who have been well served by the products and services he has helped to market. For every potential moral wrong that a person might commit, there must be an equivalent moral good that he might perform.

Those advertising agencies that have worked with the American Cancer Society to produce messages that have helped to persuade people to give up smoking or to refrain from becoming smokers have performed an important public service. Any utilitarian assessment of their performance would almost surely conclude that they had contributed to human happiness and significantly reduced the amount of pain and suffering in the world. Other advertising campaigns have assisted humanitarian organizations to raise funds for their operations and to further their causes. The government itself has employed advertising to discourage harmful behavior and to encourage beneficial activities. For example, during the Second World War, numerous ads reminded workers of the dangers inherent in talking about matters that might have helped the enemy and discouraged absenteeism at a time when the nation needed a steady supply of war matériel to carry its war effort to a successful conclusion. Similarly, advertising campaigns mounted by heavily overpopulated nations have had some effect in encouraging their citizens to employ contraceptive devices in order to bring population growth down to a manageable level. The agencies that have helped to mount such campaigns can justly take pride in their work, for it is reasonable to believe that they have contributed to the sum of human happiness through their efforts.

On the other hand, some products that are heavily advertised are known—or ought to be known by those who market them—to be dangerous and capable of inflicting grave injuries upon those who use them. For example, Ultra Sheen Permanent Creme Relaxer is an emulsion used by consumers and professional beauticians for the purpose of straightening curly hair. Ads represented it as "gentle" and "easy" to use. A woman in a television commercial for the product said that it "goes on cool while it really relaxes my hair. And the conditioner and hair dressing protects against moisture, so my hair doesn't go back." But the Federal Trade Commission found that Ultra Sheen's active ingredient was sodium hydroxide—lye—which straightens hair "by breaking down the cells of the hair shaft. . . . In some instances, [it] makes it brittle and causes partial or total hair loss." Moreover, the FTC found that it was neither cool nor gentle, but is "a primary skin irritant. It is caustic to skin and breaks down the cells which form the epidermis. Ultra Sheen relaxer in some instances causes skin and scalp irritation and burns, which may produce scars and permanent follicle damage. It also causes eye irritation and may impair vision." Since direct contact with eyes, scalp, or skin could cause irritation or injury, the FTC found that the product was not easy to use, contrary to the claims expressed in the ads. The FTC accordingly ordered the respondents to warn their customers of the product's dangers, to inform them of the presence of lye in it, to stop misrepresenting it, and to give clear instructions as to procedures to be followed in the event of injury to the customer.[3]

The law has for a long time recognized the duty a manufacturer owes to the purchaser of his product, particularly when the product is inherently dangerous. This duty has gradually been extended to others involved in the distribution and marketing of inherently dangerous products, so that persons and firms who retail automobiles, firearms, explosives, and poisons (for example) can be held liable in tort for damage caused by these products that result from defects or negligence in the way they are labeled or handled.[4] At the very least, one would expect a warning to appear on the label of any product whose use might result in serious physical injury. The creation of misleading advertisements is itself an evil that we will consider shortly. But the advertising and mass marketing of products that are prone to cause grave injury is in itself a questionable practice.

This is not to say that dangerous products, including poisons, should not be sold. In the fifth chapter of *On Liberty,* John Stuart Mill argued that people ought to be permitted to purchase poisons, but that merchants who sold such substances were under a moral obligation (which ought to be a legal obligation as well) to label them clearly so that those who purchase them will know what they are buying. Mill's label rule ought to be extended to advertisements, since the decision to purchase a product is often made soon after an advertisement is seen. The product's hazards should be prominently displayed in its ads so that the potential consumer may know what he or she is buying before the purchase is made.

2. SOME MORAL QUESTIONS ABOUT THE AUDIENCE

In morals and in law, the peculiar vulnerability of children has long been recognized. They have rightly been regarded as especially entitled to measures designed to protect them against exploitation and harm perpetrated by persons who were either unscrupulous or insensitive to their need for special treatment. It is difficult to determine whether advertisers who direct their messages to children are deliberately taking advantage of the special vulnerability of those for whom their ads are designed, or merely attempting to sell their products to the persons for whom they are intended. If they are not aware of the fact that young children are often unable to distinguish reality from fantasy, however, they have a moral obligation to study the question and to discover the facts, since their actions may properly be deemed as an exercise of power over persons who do not have the means to resist.[5]

The moral propriety of marketing dangerous products becomes even more questionable when the advertisements are aimed at children, for the likelihood that children will misuse the products or misunderstand their proper functions is far greater than the likelihood that adults will do so. The advertising of guns in Boy Scout magazines that circulate to boys from the age of twelve is of particularly dubious moral propriety. While the possession of guns by adolescents is not unlawful, there may be a distinction between what is permitted by law and what ought morally to be done or encouraged. It is easy enough to think of products and services that might not be prohibited by law but ought nevertheless to be barred from the advertising columns of magazines directed to children and adolescents.

A particularly blatant abuse of the advertiser's power to influence young minds in order to boost sales, without regard to the potential harm to the users, was perpetrated by the Hudson Pharmaceutical Corporation. The FTC prohibited Hudson from directing its advertising for "Spiderman" and other children's vitamins to child audiences. The corporation's television and comic book advertisements, directed at children under the age of twelve, used Spiderman, a cartoon hero who possesses spectacular agility and near-supernatural strength. The FTC found that the advertising in question tended "to blur for children the distinction between program content and advertising and to take advantage of the trust relationship developed between children and the program character." The ads led the children to believe that the product "has qualities and characteristics it does not have" and tended "to induce children to take excessive amounts of vitamin supplements which cause injury to their health."[6] The FTC found that Spiderman Vitamins with iron are a deadly poison when ingested in large amounts by children. In 1973, vitamins and mineral supplements accounted for more than 5,000 cases of poisoning among children under five years of age, exceeded only by aspirin, soaps and detergents, and plants (other than mushrooms and toadstools).[7]

The Hudson Vitamin case represents advertising at its moral nadir. The advertisements were directed toward children, who are particularly vulnerable, using a child's hero who endorsed an inherently dangerous product, which the children were encouraged to consume in dangerous quantities under the illusion—fostered by the ads—that they would thereby acquire the spectacular abilities of Spiderman. From a purely dollars-and-cents point of view, the campaign was eminently successful, for the sales of Spiderman Vitamins rose precipitously. But moral propriety cannot be judged by balance sheet figures.

Advertising consists of far more than the commercials that appear on television and the display pieces that are published in newspapers and magazines. Many firms prepare the potential market for their products long before the market is ready to start buying. Drug companies, for example, distribute medical instruments and other valuable items to medical students almost as soon as they enter medical school. This is done under the guise of supporting medical education. There is no need, however, for pharmaceutical manufacturers to support medical education in this fashion, by supplying students with expensive instruments engraved with their logos. Outright grants to the medical schools, which might then be used to purchase the same instruments *without* the advertising, would raise the companies above suspicion. The practice continues throughout a physician's career, with free samples of drugs and various other gifts clearly identifying the company from which they come. Unlike scientific articles in journals reporting on the efficacy of various drugs for particular purposes, such gifts are designed to impress the physician (or the budding physician) with a specific brand name, to build his loyalty to a particular company. Whatever the true intentions of their distributors may be, they can certainly be construed to be intended to override the physician's objective assessment of competing products by building loyalty to a particular firm and its products. This is subversive of the medical profession and is an unethical intrusion into a delicate process where extraneous considerations should not be permitted to interfere.

In this connection, it may be observed that large advertisers buy university professors and other researchers, who supply them with materials and opinions to "substantiate" their claims. A recent study by Representative Benjamin S. Rosenthal and the Center for Science in the Public Interest, "Feeding at the Company Trough," revealed that significant numbers of university professors of nutrition have established remunerative connections with the food and drug industries. "Unfortunately," the study says, "many professors have developed extensive ties with the same industries of which they are asked to be objective analysts. . . . Developing ties to industry causes one to overlook problems, rationalize faults, and defend policies." Despite conclusions by many respected experts in dentistry and nutrition substantiating the thesis that a diet rich in sucrose is hazardous to a person's health, one university professor who has ties with Kellogg, Nabisco, Carnation, the Cereal Institute, and the Sugar Association insists that people can double their sugar intake without risk to their health. A report prepared at Harvard under a grant from Kellogg concluded that pre-sweetened cereals have no adverse affect on tooth decay—but did not reveal Kellogg's financial involvement in the study.[8] While I would not claim that the researchers in these cases have been corrupted, their financial involvement with the companies whose products their studies have supported lends credence to the suspicion that they might have been.

Advertisers ought to adhere strictly to the following rules: that they not pitch their appeals to children, who are under the age of reason and discretion; that they not bestow gifts upon persons whose professional judgment might thereby be distorted; and that if they sponsor research into questions that might affect public acceptance of their products, their sponsorship should not be revealed to the persons doing the studies, so that objectivity may be maintained and both scholars and sponsors might remain above suspicion of corruption and connivance at a rigged conclusion.

3. THE ADVERTISER'S DUTY TO THE ULTIMATE CONSUMER

The advertising agency is hired by a firm to sell the firm's products. By signing a contract, it undertakes an obligation to do its utmost to fulfill that goal. Acceptance of that charge does not, however, relieve employees of the agency of their duties as citizens or as human beings. While their immediate goal as advertising men may be to land accounts and to keep them by increasing the customer's sales, that goal should never be achieved at the expense of harm to unsuspecting persons. The duty not to direct advertising to children is connected with this moral obligation, as is the duty to label hazardous substances with clear and unmistakable warnings. Such moral obligations may not be enforced by the law. Persons who are concerned with doing what is right need not set the limits of their conduct at the bounds delineated by the law, for the law does not always conform with standards of moral right. To be more specific, if the law permits an advertiser to refrain from mentioning a particular hazard that his product poses to his customers, it does not follow that he has a moral right to withhold that information from them. If the law provides no sanctions against deceptive advertising, an ethical advertiser will nevertheless not engage in willful deception of those who place their trust in him.

On a broad view of what advertising is, the little leaflets that are enclosed in the boxes in which drugs are packed may be regarded as advertisements of a sort. They contain technical information for the doctor's reference and are designed to prevent improper use of the drug. But they are also designed to influence physicians to use drugs for certain medical conditions, and may therefore be regarded as at least partially intended to serve as marketing devices.

A recent study revealed that certain drugs are advertised and packaged in Latin America in ways which the FDA would condemn (or has condemned) as unacceptable in the United States. Some examples:

Winstrol, a synthetic derivative of testosterone, is considered too toxic in the United States for all but the narrowest use. The AMA warns that such drugs "should not be used to stimulate growth in children who are small but otherwise normal and healthy." But in Latin America, Winstrol is widely promoted as an appetite stimulant for underweight children. A spokesman from the Winthrop Drug Company complained that the advertising was quoted out of context and that the company complied with the laws and medical practices of each country in which it did business.

Another Winthrop product, Commel (dipyrone), is a pain killer that may cause fatal blood diseases and may not be sold in the United States as a routine treatment for pain, arthritis, or fever. According to the AMA, the "only justifiable use [of dipyrone] is as a last resort to reduce fever when safer measures have failed." But a packet of the drug purchased in Brazil recommends that the drug be used for "migraine headaches, neuralgia, muscular or articular rheumatism, hepatic and renal colic, pain and fever which usually accompany grippe, angina, otitis, and sinusitis, toothache, and pain after dental extractions." The company's comments about the matter were as evasive as they were about Winstrol.

E. R. Squibb & Sons' Raudixin, which is occasionally used in the United States to treat high blood pressure, was found to induce such deep depressions that hospitalization was often necessary and suicide sometimes followed. But in Brazil, the package insert says it is the "ideal medicine for the treatment of emotional disturbances such as states of tension and anxiety, and in states characterized by nervousness, irritability, excitability and insomnia. . . . Raudixin is the drug of choice in daily practice." A company spokesman acknowledged that the insert had been written twenty years ago, conceded that it had not been rewritten since, but insisted that it complied with Brazilian drug regulations.[9]

In short, drug companies seem to feel that their principal and possibly their only responsibility is to their stockholders, and not to the persons who might use their products. The aim of their ads, and even of their package inserts, is not so much to inform physicians of the uses and potential hazards of their products as to persuade them to prescribe them. When forced to do so by government regulations, they will write truthful and informative inserts; but when government regulations are lax, they will subject the public to needless hazards and rationalize their conduct by claiming that they are doing nothing unlawful. This is not only morally unacceptable, it should also be legally proscribed. Those of us who believe in minimal government interference in private affairs would prefer to see government regulations of all industries, including the drug industry, reduced as much as possible. But so long as the industry itself behaves irresponsibly and thereby endangers the lives and health of the persons it is supposed to serve, the public has no alternative but to rely upon government for protection.

4. THE FORM OF THE ADVERTISEMENT

The ethical advertiser is obligated to assume some responsibility for the product he advertises. He ought to design his advertising campaign in such a way that it does not prey upon the gullible, the young, and the innocent. He ought to refrain from corrupting professional judgments about matters that may gravely affect important human interests. And he ought to assume some responsibility for the form of the advertisements he creates—to be certain that they do not deliberately deceive, leave false impressions, or commit fraud. Truth telling, after all, is a fundamental moral rule. While we may not agree with Kant's insistence that one never has the right to lie, whatever the circumstances, it is nevertheless true that the liar and the cheat corrupt language and destroy confidence. In their zeal for short-term profit, they encourage movements for government control over themselves and their competitors so that the weak and powerless may be protected against those who would exploit them. Freedom itself is threatened by those who cavalierly assume the right to pervert the truth for their own narrow ends. Unfortunately, advertisers have devoted much of their skill to the artful presentation of half-truths and lies, to deception of the eye and of the mind, to the creation of illusions and false expectations. These skills have proven themselves in the marketplace. False promises and clever doubletalk have been effective selling tools since the itinerant patent medicine peddler hawked his merchandise from the back of his wagon with the promise that it would cure everything from warts to cirrhosis of the liver. False promises, lies, and other deceptions *are* effective. They can make money for those who use them. They can help to drive honest competitors out of business, for it is simply not true that the truth will ultimately come out. Even if it does, it may come too late for the honest but bankrupt merchant. One immoral practice breeds another. Lies beget lies. The honest man who sees himself being pushed to the precipice of financial ruin may turn in self-defense to dishonest practices that he would otherwise abhor. The corrupt advertiser provides an excuse for every other advertiser to engage in corrupt practices: "Everyone else is doing it."

A catalogue of deceptive advertising practices would consume more space than I am permitted in this essay, but a few examples may be illuminating:

Some firms advertise a product as if it were on sale at a reduced price when, in fact, the product is never sold at the so-called regular price. A paint manufacturer, for example, advertised its paint as follows: "Buy 1 gallon for $6.98 and get a second gallon free." But it *never* sold its paint at $6.98 per gallon!

Encyclopedia salesmen misrepresent themselves as agents for school boards or as public opinion pollsters. They offer "free" sets of their encyclopedias, allegedly as a public relations "service," but ask for a small monthly service charge for a ten-year research service which will presumably guarantee the worried customer's children their places in medical school or law school.

Record clubs and book clubs falsely advertise "free" books which are not free at all, but are consideration for a binding contract to purchase a number of books at a supposedly reduced price which turns out, after postage and handling charges are added, to be higher than the retail price of the same books. The clubs are of course not clubs at all, since their members never meet. The members are customers and the clubs are profit-making businesses.

Insurance companies and other firms utilize photographs or drawings of immense buildings in their ads and on their stationery to suggest that they are large, long-established firms, even though they may occupy no more than a single office. One firm recently hung its own sign on a large, modern municipal government building and filmed its commercials in front of the disguised structure, leaving the impression that its own offices were housed there.

Small type may be used to obscure limitations on insurance coverage, while bold type may emphasize irrelevant facts, such as coverage which is common to *all* policies of a given class—the old fallacy of accent.

Misleading words and phrases—particularly those having special technical meanings that are unfamiliar to the uninitiated—are used to create false impressions in the minds of laymen. Ordinary language may be used in such a way as to suggest to the uninitiated that certain conditions apply when they do not apply at all. For example, an ad saying "This policy will pay your hospital and surgical bills" suggests (though it does not literally say) that *all* of the hospital and surgical bills of the insured will be paid; and "This policy will replace your income" suggests that *all* of the insured's income will be replaced if he becomes disabled—when, in fact, only a very small portion of his bills or his income will be paid or replaced.

Some companies don't hesitate to make inconsistent claims for competing products that they manufacture or distribute. The Sterling Drug Company, for example, distributes Bayer Aspirin in the United States and also manufactures "Vanquish." In 1970, the company was simultaneously running ads which made the following claims:

For Bayer: "Aspirin is already the strongest pain reliever you can buy." Combining Bayer with other drugs or buffering it would not improve it. "No one has ever found a way to improve on Bayer Aspirin, because Bayer Aspirin is 100% Aspirin. None is faster or more effective than Bayer Aspirin. Even WE can't improve it, though we keep trying."

For Vanquish: It has "a unique way" of relieving headache "with extra strength and gentle buffers. . . . It's the only leading pain reliever you can buy that does."

Thus, Sterling Drug is both unable to produce a pain reliever that is more effective than Bayer, and it is. Buffering doesn't add to the strength or gentleness of aspirin, and it does. Remarkable![10]

Another device is the half-truth, which becomes an outright lie because it creates a completely false impression. Excedrin, for example, ran an ad reporting that a "major hospital study" showed that "it took more than twice as many aspirin tablets to give the same pain relief as two Excedrin." But the ad failed to point out that the Excedrin tablets contained twice as much aspirin as plain aspirin tablets and that another study had demonstrated that Excedrin had caused more intestinal upset than two brands of aspirin when given in equal doses.[11]

While most of these examples have been derived from studies of the drug industry, that industry has no monopoly on deceptive advertising practices. Similar examples can be cited from industries as diverse as automobiles, lumber, oil, household cleaning products, and real estate. Immoral and irresponsible persons and corporations in these and other industries have caused severe financial losses to unsuspecting individuals who relied upon what they were told or thought they were told by hucksters whose principal concern was the closing of a sale. Even where the individual's loss is relatively small, collectively the damage may amount to hundreds of millions of dollars. With these financial losses, there are inevitably other costs that are more difficult to assess, not the least of which is the emotional damage, the anger, and the resentment that must follow in those cases where the loss represents a major portion of an individual's earnings or savings. Some of this undoubtedly spills over into resentment against a system that permits what the victims perceive to be grave injustices against themselves and the classes or groups to which they belong. While advertising alone cannot be held responsible for any social dislocations that might result from such resentments, the advertising industry cannot wholly escape responsibility for its contributions to the sense of injustice that prevails in so many quarters today.

In sum, admen should no longer allow themselves the luxury of supposing that anything goes so long as it works, or of using as the criterion of "bad" advertising practices the likelihood of failure or the possibility of governmental or consumer backlashes that might ultimately result in a loss of sales. As citizens, they have moral and social responsibilities that transcend their narrow business interests. The sale of worthless products or services, or even worse, of harmful or dangerous products without adequate disclosure of the risks; the corruption of science and the professions; the deliberate spread of lies and deceptions—all of these are contrary to the principles of decency, morality, fairness, and justice, upon which a democratic society depends for its existence. They constitute a kind of subversion, a menace to the public welfare, one that cannot be tolerated in a society that aspires to be free and just. The elementary moral principles of truth, fair play, and decency toward one's fellow man are the foundations of a free society. It cannot be in the long-term interests of businessmen to violate those principles.

NOTES

1. Cf. Deut. 19:14; *Baba Meziah* 4:12, 60a–b; *Baba Batrah* 21b, and elsewhere in both the Bible and the Talmud. Cf. *Encyclopaedia Judaica* vol. 7, s.v. *Hassagat Gevul,* pp. 1460 ff.
2. Bruce Barton, "Advertising: Its Contribution to the American Way of Life," *Reader's Digest,* Vol. 66 (April, 1955), pp. 103 ff.
3. FTC in the Matter of Johnson Products Company, Inc., and Bozell & Jacobs, Inc., Docket no. C-2788 (February 10, 1976).
4. Cf. John G. Fleming, *The Law of Torts,* 4th ed. (Sydney, Australia: The Law Book Company, Ltd., 1971), pp. 452 ff.
5. See *ITT Continental Baking Co., Inc., et al. Final Order to Cease and Desist, and Opinion of the Commission* (Wonder Bread Case), Dkt. 8860 October 10, 1973, 3 Trade Reg. Rep. Par. 20, 464 at 20, 372, January 14, 1974, cited in *Action for Children's Television Petition before the FTC,* October, 1975 (hereinafter referred to as *ACT*), p. 22.
6. FTC, File no. 762 3054, *Consent Order to Cease and Desist and Accompanying Complaint.* September 2, 1976.
7. *ACT,* Appendix G, Tables 1 and 2.
8. *ACT News,* Vol. 6 (Fall, 1976), p. 11.
9. Cf. Robert J. Ledogar, *Hungry for Profits: United States Multinationals in Latin America* (New York: IDOC/North America, Inc., 1976).
10. From Select Committee on Small Business of the U.S. Senate (92nd Congress), Hearings before the Subcommittee on Monopoly on Effect of Promotion of Advertising of Over-the-Counter-Drugs, Part 1, Analgesics (1971), p. 230.
11. Ibid.

11

ADVERTISING AND ETHICS

PHILLIP NELSON

THE MARKET SYSTEM AND ADVERTISING

There are two possible routes one can take to ethics. One can exhort others to take account of social well-being in their behavior—"to love one another" and act accordingly. Or one can try to design institutions such that people will, indeed, benefit society, given the motivations that presently impel their behavior. Most economists, whatever their political position, adopt the latter view; ethical behavior is behavior that, in fact, benefits society, not necessarily behavior that is motivated to benefit society.

Those of us who advocate the market as an appropriate institution are following the lead of Adam Smith: that the market, more or less, acts as if there were an invisible hand, converting individual actions motivated by the pursuit of private gain into social benefit. The selfish employer, for example, callously firing employees when he no longer needs them, helps in the reallocation of labor to activities where it is more useful.

This is not the stuff of poetry. In novels—and quite possibly in the interpersonal relations upon which novels generally focus—selfish people act in ways disastrous to those around them. But novels are hardly the basis for determining social policy, though novelists and their compatriots, literary critics, are often in the forefront in the espousal of "social causes." They have been the consciences of society. Because of their focus on motivation, they have generated a guilt complex when guilt is totally unjustified.

It must be admitted that the market is not a perfect instrument, that the invisible hand wavers a bit. Some individual actions will not lead to social well-being. However, popular perceptions tend to exaggerate market imperfections. For example, the available evidence indicates that the monopoly problem is not terribly serious in the United States.

More importantly, the popular view fails to evaluate the problems of alternative institutions. The record of government regulation to make the market behave has been distinguished by case after case where the cure has been worse than the disease, where often there has been no disease at all.

I want to look at the ethics of advertising, given this perspective. Advertising is ethical not because of the motivations of its practitioners but because of the consequences of its operation. The invisible hand strikes again! The market power of consumers will force advertisers to act in ways that benefit society. Advertising will by no means be an "ideal" institution. But it will do an effective job of getting information to consumers.

Advertising bothers its critics not only because its practitioners are selfishly motivated. The advertising itself is often distasteful. Celebrities endorse that brand that pays them the highest price. Advertisers lie if it pays. Advertisers often make empty statements. Nobel prizes for literature have not yet been awarded to the classics of the advertising art. But the crucial question is not whether advertising is aesthetically satisfying, or whether its practitioners are noble, or even whether they occasionally lie. The question is whether advertising generates social well-being. Some of the former questions are not irrelevant in determining the answer to the latter question. In particular, as I discuss later, the role that truth plays in generating socially useful advertising is an important question.

ADVERTISING AND INFORMATION

Before resolving the fundamental ethical issue about advertising, it is important to understand how advertising behaves. I support a simple proposition about the behavior of advertising: that all advertising is information. This is not a statement with which the critics of advertising would agree. What bothers them is that advertising is paid for by the manufacturers of the brands whose products are being extolled. How can information be generated by such a process? Clearly, some kind of mechanism is required to make the self-interested statements of manufacturers generate information. But such a mechanism exists—consumer power in the product markets.

The nature of consumer control over advertising varies with the character of consumer information. Consumers can get some information about certain qualities of products prior to purchase. For example, they can try on a dress, find out about the price of the product, or see how new furniture looks. I call these "search qualities." In the case of search qualities, a manufacturer is almost required by the nature of his business to tell the truth. The consumer can determine before he buys the product whether indeed this is the dress or the piece of furniture that has been advertised; and in consequence, it will pay the advertiser to be truthful. This is a situation where the famous ditty of Gilbert and Sullivan would be appropriate:

> This haughty youth, he tells the truth
> Whenever he finds it pays;
> And in this case, it all took place
> Exactly like he says.

Now, there are other qualities that the consumer cannot determine prior to purchase. It is very difficult for the consumer to determine the taste of a brand of tuna fish before he buys the tuna fish, or to determine how durable a car will be until he's experienced it; but even in these cases, the consumer can get information about a product. The character of his experiences when using the brand will generate information to the consumer. This information will not be useful for initial purchases, but it will govern whether the consumer repeatedly purchases the brand or not. The repeat purchase of consumers provides the basis of consumer control of the market in the case of "experience qualities."

In this case, there will be certain characteristics of the advertising which are truthful. It will pay the advertiser to relate correctly the function of the brand. It pays the manufacturer of Pepto-Bismol to advertise his brand as a stomach remedy rather than as a cure for athlete's foot because, obviously, he is going to be able to get repeat purchases if Pepto-Bismol does something for stomachs and people are taking it for stomachs. If they're taking the stomach remedy for athlete's foot, they're in trouble. So the effort to get repeat purchasers will generate a lot of truthfulness in advertising. Another example: it pays the manufacturer of unsweetened grapefruit juice to advertise the product as unsweetened. This is the effective way to get repeat purchases; hence people can believe it.

There are other qualities about the brand for which the incentive of truthfulness does not exist. It pays the manufacturer of Pepto-Bismol to advertise his brand as the most soothing stomach remedy even if it were the least soothing stomach remedy around. It pays somebody to say that a piece of candy tastes best even if the candy has an unpleasant taste. Even here, however, there is information for the consumer to obtain through advertising. The advertising message is not credible, but the fact that the brand is advertised is a valuable piece of information to the consumer. The consumer rightly believes that there is a positive association between advertising and the better buy. The more advertising he sees of the product, the more confidence he has prior to purchasing the product. Simply put, it pays to advertise winners. It does not pay to advertise losers. In consequence, the brands that are advertised the most heavily are more likely to be the winners.

The mechanism that is operating is the repeat purchasing power of the consumer. Brands that are good after purchase will be brands that consumers buy more. In consequence, there is a negatively sloped demand curve. People buy more as the price per unit of utility of a good goes down, even when it takes experience on the part of consumers to determine this utility. As quantity goes up, the amount of advertising will also go up. This is a well-established relationship.[1] The positive association between quantity and advertising and the negative association between quantity and price per unit of utility generates a negative association between advertising and price per unit of utility. In other words, the "better buys" advertise more; and, in consequence, the *amount* of advertising provides information to consumers.

Considering that we have no direct measure of "better buys," there is a good deal of evidence to support this proposition. First, it pays a firm to expand its sales if it can produce what consumers want more cheaply than other firms. It can increase its sales either by increasing advertising or lowering prices. I maintain that it does both at the same time, just as plants usually increase both their capital and their labor when they expand output on a permanent basis. But the critics say that the larger selling brand advertises more; therefore, it charges more to cover the costs of advertising.

The only way the critics could be right is if diminishing returns in advertising did not exist. By diminishing returns in advertising, I mean that the more a manufacturer spends on advertising, the less he gets in additional sales per dollar of advertising. When there are diminishing returns, the advertising of the larger selling brand is less efficient; it gets fewer sales per dollar. When advertising is less efficient in that sense, the larger advertiser will have a greater incentive to get additional sales by lowering the price. With diminishing returns in advertising, then, the larger selling brand both advertises more and gives greater value per dollar. There is considerable evidence that there are, indeed, diminishing returns in advertising.[2]

There is a second strand of evidence in support of my position. One can successfully predict which products get advertised most intensively by assuming that advertising provides information in the way I have described. It can be shown that it requires more advertising to provide the indirect advertising for experience goods than the direct advertising of search goods. Indeed, the advertising/sales ratios are greater for experience goods than search goods.

There is another important piece of evidence that winners are advertised more. If it is true that the larger-selling brand provides better value per dollar on the average than smaller-selling brands, wouldn't it pay a brand to advertise its rank in its product class more, the higher the rank? Consumers would prefer to buy top sellers rather than bottom sellers. The evidence is overwhelming that more brands say that they are Number One than declare any other rank.

One could argue, I suppose, that consumers are brainwashed into believing that larger-selling brands are better, when the contrary is true. But how could this be? A lot more advertisers have an interest in brainwashing the consumers into believing the contrary. Yet, the "big is beautiful" message wins. The only reasonable explanation is that this is the message which is confirmed by the consumers' own experiences. The brainwashing explanation is particularly hard to accept, given the industries in which brands most frequently advertise their Number One status. It pays consumers to make much more thoughtful decisions about durables than non-durables because the cost to them of making a mistake is so much greater in that case. Yet, the "I am Number One" advertising occurs more frequently for durables than for non-durables.[3] Even more convincing is the evidence that the advertising of Number One rank is not confined to possibly gullible consumers. That same message is used in advertising directed to businessmen. They too must have been brainwashed if the critics are right. But such soft-headed businessmen could hardly survive in the market.

The evidence seems inescapable: larger-selling brands do, on the whole, provide the better value per dollar. The evidence also shows—and all would admit—that larger-selling brands advertise more. In consequence, the more advertised brands are likely to be the better buys.

It is frequently alleged that advertised brands are really no better than non-advertised brands. A case that is often cited in this connection is Bayer Aspirin. But aspirins do, indeed, vary in their physical characteristics. Soft aspirins dissolve in the stomach both more rapidly and more certainly than hard aspirins. In consequence, the soft aspirin are better. They are also more expensive to produce. It is no accident that the most heavily advertised brand of aspirin is a soft aspirin. Of course, there are also non-advertised soft aspirins that sell for less than Bayer Aspirin. But the issue is not whether the best unadvertised aspirin is as good as the most heavily advertised aspirin. The issue is whether purchasing one of the more heavily advertised aspirins at random gives one a better product, on the average, than getting an unadvertised aspirin at random. The existence of unadvertised soft aspirin, when the consumer does not know which aspirin fits into that category, is of little help to the consumer.

Advertising can provide this information without consumers being aware of its doing so. Advertising as information does not require intelligent consumer response to advertising, though it provides a basis for such intelligent response. Consumers who actually believe paid endorsements are the victims of the most benign form of deception. They are deceived into doing what they should do anyhow.

It does not pay consumers to make very thoughtful decisions about advertising. They can respond to advertising for the most ridiculous, explicit reasons and still do what they would have done if they made the most careful judgments about their behavior. "Irrationality" is rational if it is cost-free.

Whatever their explicit reasons, consumers' ultimate reason for responding to advertising is their self-interest in so doing. That is, it is no mere coincidence that thoughtful and unthoughtful judgments lead to the same behavior. If it were not in consumers' self-interest to respond to advertising, they would no longer pay attention to advertising. It is in this context that we can examine the question of whether true or false advertising statements should be protected by the right of free speech.

DECEPTIVE ADVERTISING, REGULATION, AND FREE SPEECH

A lot of people think that there is far too much deceptive advertising and that active government intervention, policing every advertisement, is necessary to improve advertising.

But the amount of deceptiveness in advertising can be easily exaggerated if one simply looks at the incentives of the advertiser to deceive without considering the incentives of consumers not to be deceived. The circumstances under which advertisers have the greatest incentives to deceive if consumers believed them are precisely the circumstances under which consumers would be least inclined to believe advertising. Deception requires not only a misleading or untrue statement but somebody ready to be misled by that statement.

One possible source of deceptive advertising is consumer confusion. Though the decision rules that consumers need follow to avoid being deceived by advertising are relatively simple, some consumers will possibly be confused. They will possibly be gullible when they should not be, and inappropriately skeptical at other times.

But let us suppose that one tries to remove these deceptions by active government intervention to prohibit advertising that deceived anybody. Short of eliminating all advertising, such a government role would be self-defeating. Whatever the standards for fraudulence in advertising, it is unlikely that all consumers will know those standards. If the relatively simple decision rules necessary to avoid deception without government intervention are too confusing to some consumers, how much more confusing is the law on fraud? I know of no simple (or complicated) decision rule that would tell a consumer which advertising claims are legally required to be valid and which are not, or which advertisements are legally misleading and which are not.

The more the law protects against fraud, the more people think the law protects against fraud. Misinterpretation of the law's domain will exist, no matter how extensive that domain. Indeed, I believe, there is probably more deceptive advertising when laws on fraud exist than when they do not. Consumer market power generates information from advertising precisely because that information is in the self-interest of producers to provide. Hence, there is little incentive for deceptive advertising under the aegis of consumer market power. In contrast, state police power, involving the expenditure of resources, will never be enforced vigorously enough to eliminate all incentives for fraudulent advertising even in terms of the legal definition of fraud that prevails at the moment.

I am not saying that these laws against deceptive advertising are pointless. I am only asserting that most people have missed their point. The virtue of these laws is not that they reduce deceptive advertising. Rather, it is that they can make more information available to consumers than they would otherwise receive. Take, for example, the law prohibiting the mislabeling of the fabric content of clothing. If that law is sufficiently enforced, consumers will believe that a clothing label is usually correct. This will provide an incentive for some manufacturer to mislabel—unless the law is enforced so vigorously that nobody gains from breaking it (a non-optimal level of law enforcement). In the absence of the law, no one could trust any clothing label that it was not in the self-interest of the producer to specify correctly. Hence, these clothing labels, though incorrect, would not deceive many people. This law is not reducing deception in advertising, but it is enabling consumers to determine in many instances the fabric content of their clothing from the label. Laws can achieve the objective of more direct information at the price of both enforcement costs and costs to the consumer of being deceived where otherwise he would be appropriately distrustful.

Laws against fradulent advertising are trying to accomplish something very important. People who feel that advertising is often wasteful have some real basis for that feeling. In the same sense that the engineer rates some kinds of engines as inefficient, the economist can declare that advertising is inefficient, that there is a significant potential for the improvement of advertising's performance. Whether that potential can be realized economically or not is, of course, another question. An inefficient motor may be the best motor we have. The indirect information which dominated the advertising of experience goods is inefficient compared to direct information. (By indirect information, I mean the information contained in the *fact* that the brand advertises. This is in contrast to direct information, which is the specific information contained in the advertising message.) First, direct information tells the consumer more, as evidenced by his preference for direct information. Second, more advertising is required to transmit indirect information than direct information. Any increase in the proportion of direct information in the advertising of experience goods would be an increase in advertising's efficiency. Laws that increase the range of believable statements in advertising can help advertising do a better job—unless, of course, they create even worse problems.

There are, indeed, some serious drawbacks to the way in which the Federal Trade Commission has been enforcing the law against fradulent advertising. Hyperbole plays a useful role in advertising. Exaggeration makes advertising more memorable. The more memorable the advertising, the more efficient it will be from both a private and social point of view, simply because memorability makes advertising perform its information function better. The Federal Trade Commission seems bent on eliminating these exaggerations. Take, for example, the famous case of the sandpaper shave. The FTC ruled that Rapid Shave must cancel this advertisement on the grounds that the conditions of the experiment were not quite kosher. An exceedingly memorable advertisement was eliminated. I believe virtually no new source of direct information was created by this decision.

Take another case on which the FTC would be on stronger ground: advertisements comparing Shell's performances with and without TCP, a gasoline additive. Obviously, it is an irrelevant comparison because Shell without TCP is not an option facing consumers. What harm is done by eliminating this advertisement? Its very existence suggests that it is a memorable advertisement. What good is done by this act? Nothing, by itself. The FTC would have to police carefully all advertisements using any purported tests or surveys to determine both the relevance and the appropriateness of the study design. Even then, it would take consumers quite a while to begin to believe the data quoted by advertisers, because it has been in their interest to distrust these data for so long. This herculean effort by the FTC would, in the process, have the unfortunate by-product of eliminating lots of memorable advertising.

AN ALTERNATIVE PROPOSAL

Can we do better? Laws designed to prevent deceptive advertising are not necessarily the best laws to help open up new bases of believable statements in advertising. The more strictly laws on fraud are interpreted, the less the opportunity for hyperbole, with its very real information payoff to society.

I think it is possible to design laws that at the same time: (1) create new bases for direct information, (2) allow hyperbole as much sway as the market desires, and (3) reduce enforcement costs as compared with the present legal structure. Remember, that is the heart of the problem. Advertisers provide indirect information for experience goods because, in the absence of laws, they are limited in the kinds of direct information they can authenticate. Let us attack that problem in the most direct way possible.

Let us create a dual set of standards for advertising authenticity. Suppose the advertiser were to utter some magic words such as, "We guarantee the validity and relevance of the information contained in this commercial." Then he would be strictly accountable in the courts for that information. Not only must that information be true, but it must not be misleading. Without such magic words, the advertiser is not accountable. Such a legal structure accomplishes the same objective as government authentication of advertising messages. The only differences are that the authentication process occurs after the advertisement rather than before, and the courts are the functioning government agency rather than the FTC.

This system seems to me to be vastly preferable to the present vigorous policing policies of the FTC against purportedly misleading advertising. It permits the authentication of direct information without destroying the effectiveness of indirect information. Certainly, the criterion that the FTC uses in policing advertising—does it mislead any consumers?—seems a self-defeating criterion to employ. The very act of policing makes other advertisements misleading to some consumers who would not have been previously misled.

It seems to me that, worthy though it may be, the goal of converting the advertising of experience qualities from indirect information cannot be completely achieved. Therefore, one has to remember that indirect information still has a role to play. One must be on guard against destroying the effectiveness of this indirect information while pursuing the goal of improving the information content of advertising.

In that sense, I am in favor of virtue rather than vice. But vice is not utterly vicious in its consequences. Consumers are, on the whole, better off as a result of being exposed to the essentially empty messages of experience goods advertising. That advertisements contain all sorts of non-credible statements does not prevent these advertisements from serving a social purpose, though that purpose could be better served if more credible statements were part of advertising.

Many of those who attack advertising are really attacking the market system in general. "Advertisers," they say, "are constantly pushing products which consumers either do not need or are positively harmful to them. For shame!" But these are always products that the consumer wants. Otherwise, the advertising could not succeed. In consequence, this is not an attack on advertising so much as an attack on consumer free choice.

It seems to me that consumer choice is the best way we have to determine what is valuable to consumers. Admittedly, consumers will make many mistakes in their choices. The information shortage that confronts consumers in their choice of brands is matched by the information shortage they face in choosing products. But consumers have a strong incentive to try to make the best decision, since their own well-being is at stake.

On the whole, government agencies will have even more severe information problems about which products are valuable to consumers than will consumers themselves. Consumers will, in general, know more about their own idiosyncracies than will any regulatory agency. The only class of cases in which the government could know more is that in which individual differences among consumers are largely irrelevant—not a very large class of cases.

While it is conceivable that in certain situations regulatory agencies will know more than will consumers, it is inconceivable that regulatory agencies will have as strong incentives to make the right choice as will consumers. Government officials are, on the whole, interested in their own well-being. Their own well-being is not closely related to consumer well-being because the voting market is a seriously defective market. People in general will have less incentive to acquire information as voters than they do as consumers, because the individual returns to them of voter information are less than the returns to them as individuals of consumer information. In consequence, government incentives to make decisions maximizing the well-being of voters are generally less than the incentives of consumers to achieve the same objective. In consequence, government decisions are likely to be worse than those of consumers.

The record of government regulation of consumer decisions is not one to encourage the case for government interference with consumer choice. Peltzman, for example, finds that the regulation of drugs by the Federal Drug Administration has made consumers far worse than they would have been otherwise.[4]

As far as advertising is concerned, the attack on consumer choice is particularly pointless. There is nothing in the process of advertising that makes advertisers systematically advertise the products that are bad for consumers rather than the products that are good for consumers. However, there is a process that seems to generate that result.

Advertising volume will be concentrated in the advertising of products that most consumers want rather than the products that satisfy minority taste. Best-sellers will be advertised more than poetry. Most critics of consumer choice would like consumers to have the tastes developed by a combined Ph.D. in English, art history, and music. Even supposing that Mozart provides more enjoyment than Elvis Presley, if one has developed the taste for Mozart, it is not clear whether such investment pays. It is assuredly clear that providing Mozart for people who prefer Presley will not work, since the Presleyites simply will not listen. The critics mistakenly attack advertising as catering to the products that are popular rather than unpopular. They are mistaken not so much in their facts as in the inference they draw from this feature of advertising.

Furthermore, advertising is in many ways more useful to minor segments of the market than to majority groups. The continuation of activities for minor segments of the market is more dependent on advertising than is the continuation of majority activities. The smaller the group, the more difficult is the information problem of matching audience and activity. Anything which reduces that information problem—as does advertising—will tend, therefore, to be of particular benefit to minority groups.

While a critic (if pressed) might admit all of the above, he could still assert that an ethical advertiser should refuse to advertise a product which is not useful to consumers. This would be a largely pointless protest upon the part of advertisers unless many advertisers operated in this fashion since, otherwise, the consequence would be that somebody else, of roughly the same competence, would do the job at roughly the same price. If enough advertisers behaved this way, the price of "unethical" advertising would go up to some extent; but given actual behavior, the ethical advertiser would be making little social contribution.

It is even questionable whether advertising dominated by this much love of humanity is desirable or not. It is a maxim that, in a full-information world, the more people love one another, the better. (Becker's work on marriage and social interaction bears this out.[5]) But with less than full information, this proposition no longer necessarily holds. One of the great advantages of selfish behavior in a market system is that it requires limited information to do a good job. The advertiser simply has to know how to advertise and at what price his services are demanded. But, now, let us convert the advertiser to a philanthropist, to a man who insists that he will advertise only those products which are best for consumers. To do his job right, he must now determine what are the consequences to consumers of all the alternative products that they face. The investments in necessary information required would be enormous, and a great deal of social waste would result. I would also be afraid of the results. The decisions people make about what others ought to consume differ in systematic ways from the decisions they make about what they themselves want. The fun in life tends to be eliminated in decisions for others. Sermons would be advertised heavily. Candy not at all. There are advantages to the maxim: Advertisers, stick to your copy.

NOTES

1. This is borne out by data from the Internal Revenue Service *Source Book of Income,* 1957. For every industry, firms with larger sales advertise more.
2. See Phillip Nelson, "Advertising as Information Once More" (forthcoming).
3. In the May, 1955, issue of *Life* magazine, there were twelve durable and three non-durable "I am Number One" advertisements.
4. Sam Peltzman, "An Evaluation of Consumer Protection Legislation: the 1962 Drug Amendments," *Journal of Political Economy,* Vol. 81, No. 5 (September, 1973), pp. 1049–1051.
5. Gary Becker, "A Theory of Social Interaction," *Journal of Political Economy,* Vol. 82 (November, 1974), pp. 1063–1093.

VI

RESPONSIBILITY: The Individual, The Organization, The Profession

12

"I ONLY WORK HERE":

Mediation and Irresponsibility

JOHN LACHS

The countries of Eastern Europe are said to be notorious for worker irresponsibility. There, production is inefficient, the performance of duties on a job is perfunctory, the quality of service is inferior. Few of the workers display interest in their work; fewer still take any pride in what they do. Travelers report that censure from supervisors is greeted with sullenness, complaints from customers with a shrug of the shoulders or a quiet curse. On being called on the carpet, the standard defense for incompetence or for anything that goes awry is the declaration "I only work here. Why pick on me?"

"I only work here" implies that I do not make the rules and do not wish to be responsible for them. It suggests that I have little notion of the interconnectedness of things and desire to restrict my operation to the narrowest limits of my job. This is one of the meanings, in fact, of the word "job." A job, in this sense, is contrasted with a position or a profession; it is simply a task to be performed for pay in as mechanical a fashion as possible. Individuals who hold down jobs in this sense make sure that there is little personal involvement in their work. To them, work is something that must be done for a living. Their conception of it is expressed accurately by the description a telephone company employee once gave me. "It is as if I took a deep breath at 8:15 in the morning and went under water," he said. "I surface about 5:00 in the afternoon and breathe a sigh of relief."

Jobs, in the current sense, are narrow roles in which virtually each operation to be performed requires explicit instructions. It is as if the human beings who perform these roles were denuded of activity and waited like machine parts to be caused to act. Since the cause of the action, the purpose, the motive, all come from the outside, it is not altogether surprising that people fail to feel responsible for failures. Machines, after all, are mere instruments, and no one could rightly blame the tires for crushing a man when the driver of the car is drunk.

There is something reassuring about the thought that such irresponsibility is rampant in Eastern Europe. The natural corollary is the supposition that it cannot happen here. Yet employers and customers alike are keenly aware that something quite like this is a widespread and growing reality in our own country, as well. I hear businessmen complain vigorously about the difficulty of obtaining the services of people who are willing to work. And what they mean by "someone willing to work" is someone who is self-motivated. That, in turn, involves the readiness to assume responsibility for getting things done, along with the corresponding responsibility for failure when that occurs.

Worker attitudes are studied in detail by industrial psychologists and by sociologists. Yet we do not have a conceptual framework for understanding why narrow job mentality spreads irresistibly, like a disease that threatens to disable the efficiency and the responsiveness of American business and industry. There is some talk about worker disaffection, about alienation, about unhappiness. But few have managed to find an intelligible pattern to it all, and fewer still have practical recommendations.

I believe that it is possible to develop concepts that help to explain the source of the irresponsibility in the face of which we manage and in the teeth of which we must live. Before I talk about the concepts that are adequate to do the job, however, I want to reject some other explanations. I must confess to having very little in the way of a precise idea when people talk about "alienation." I would readily attribute this to my own ignorance, were it not for my belief that those who talk about alienation do not really have any very precise thought in their heads either. It is evident that to call someone alienated is to express reproach or regret: it is the sort of charge that on a personal level either roils the blood or evokes winces for sympathy. To be alienated is to be in a situation that is *bad.* We may not know why or how things got this way. Most frequently, "alienation" is a term that designates some indeterminate objective evil, along with some very specific feelings of misery.

If there is no help from this side, there is even less from the side of those who claim that the key to achieving worker responsibility is to have them share more equally in the fruits of production. Inequality, injustice, and want are not major sources of negligence. By the same token, material wealth does little to render people more conscientious. On the contrary, the more we think of prosperity as our birthright, the less we understand its relation to a responsible role in the productive process. The result is that people demand to have all manner of good things, while with a clear conscience they offer little or nothing in return. I must dismiss, therefore, the sometimes exaggerated claim of unions that if only workers were treated better, earned more money, and received greater fringe benefits, they would take greater interest in their work and assume responsibility for doing well what must be accomplished.

Management frequently responds to the union claim by the classic capitalist argument that it is not higher wages that create responsibility in a corporation but actual ownership. There is some truth to this idea. Generally, we tend to be more responsible in handling things that belong to us than in dealing with objects or situations where success or failure makes little difference. But the interpretation of ownership and its embodiment in the form of holding shares, both of which are integral parts of this analysis, are inadequate. I shall try to show that economic ownership must be only a part of a larger appropriation by workers of the activities of the corporation as their own. The fundamental notion of ownership is not economic, therefore. It is personal and psychological, and requires an understanding of the organic interconnectedness of the operations that constitute a company. To own five shares out of fifty million outstanding makes little difference by itself. The very idea that the paper certificate one gets from the transfer agent in fact makes one a part owner of a company is difficult to conceive. The effectiveness of the thought that I am an owner of the company is a direct function of the understanding I have of the company and of the strength of my belief that its operations are somehow mine. If this understanding and this belief are lacking, the thought that we are owners becomes empty: its only content is the expectation of the dividend.

I want, then, to reject the notion that the primary source of our troubles is economic. In fact, I want to repudiate any thought that would identify the problem as due to some malfunction of society. Our intellectual history is a record of violent disagreements about what special structure or feature of modern industrial civilization is the primary cause of the depersonalized, irresponsible, alienated state in which we live. It is time to undercut this interminable debate by what I think is a radical departure. My argument is that no special structure is responsible for where we are. Although there may be malfunctions in our society, these cannot adequately account for our maladies. The narrow job orientation of employees, the social irresponsibility of employers, the general disaffection and depersonalization we all experience are natural costs of large-scale human society. When I say natural costs, I do not mean costs that cannot be minimized. My point is just that any society in which there is large-scale mediation will unavoidably tend to render individuals passive, ignorant, and irresponsible.

Mediation is a universal phenomenon, although it is rarely called by that name. It is simply the performance of an action by some agent on behalf of another. If you ask me to bring you a glass of water and I do, the action is mediated. It is mediated because I did it for you. Had I not done it, you would have had to do it yourself; in some sense, therefore, I performed an action that was or was to be yours. The same is true when there is no specific request for the performance of the act. I never ask the people at the generating plant to make me some electricity, yet they do it on my behalf. If they did not provide the power to turn on my lights, I would have to do it myself or make some alternative arrangement. That they are working on my behalf is amply demonstrated by the fact that I pay them.

The contracts with distant people to perform actions on my behalf can be more open-ended and informal even than this. When Frank Sinatra cuts a new record, nothing is further from his thoughts than that some obscure philosopher in Tennessee will use it to express his love. Yet Sinatra's singing captured on disk can fulfill just that function, and it is not too farfetched to say that the series of acts of which it consists were performed for (among others) me. In this way, actions that range from the trivial to the most momentous, from the specifically contracted all the way to the generally available, are all mediated—that is, performed by one on behalf of and frequently for the benefit of others. Almost all the actions necessary for life, for satisfaction, and for self-expression are mediated in industrial society.

Mediation has four major consequences. The first is impressively beneficial. It is, simply put, to make civilization possible. For the division of labor is simply a form of mediation, and mediation—at its best—is just cooperative effort. If each person had to do everything for himself, the world would fall apart into a string of isolated and, I hasten to add, starving hermits. What anthropologists call the benefits of culture are simply the record of past acts done on our behalf (or at least turned to our benefit) by people who act as our agents. Teachers and engineers, businessmen and bulldozer operators all do their jobs, perhaps unmindful of what these mean to *me*. Yet they enable me and others to do what we want and to live in the way we will.

We all know the benefits of mediation. I omit a full discussion of them not because I think them insignificant but precisely because they are uncontroversial and evident. We are so blinded by the good, in fact, that we altogether overlook the cost. If I dwell on the costs, therefore, it is not to minimize the benefits, only to present a more balanced picture.

The three remaining major effects of mediation are all costs. The first is the growing manipulation of people. This is natural, given the fact that in mediation others perform my acts for me. Since I want only the act and its desired consequences, my interest is in bringing it about quickly, certainly, and at the least possible cost. The person who is to perform it becomes an instrument of my will; my interest in him is limited to the role he performs. He is not a person, or at least not a person primarily, in my eyes. He is a means to the end I desire, and I must be sure that I know what to say, do, or promise to elicit the desired response. When such manipulation becomes necessary on a large scale, virtually everyone begins to feel used and retaliates by using others. Obviously, people do not become things simply because of this. But the more we use people, the more we adopt toward them the attitude we normally have only toward inanimate physical objects. Depersonalization consists in the disregard of our subjective, emotional inner life in favor of exclusive focus on eliciting desirable behaviors.

The second deleterious result of mediation is our growing sense of passivity. Roles in social institutions are defined functionally: their usefulness and justification derive largely from the body of mediated acts they demand. The more others look to us for the satisfaction of their desires, the more our actions must be reliable, predictable, sane. The natural tendency is to surrender oneself to one's role. If we do what the rules require without thought and without fail, we will have fulfilled, we believe, all our responsibilities.

Not surprisingly, the more we make ourselves parts of an extended chain of mediation and the more we identify with our roles, the more we surrender any claim to be the causes of our actions. A sense of passivity then descends on us like dusk; in reflective moments, we begin to feel that even the actions we physically perform are not really our own. And there is some truth to this frustrating suspicion. It is not that the industrial world leaves us passive or lazy. On the contrary, we are busier than ever, doing for others what they cannot or will not do for themselves. It is just that the motives for our acts lie not within us; what we do is not a natural consequence of our dispositions or choices. It is as if there were only one decision to make: to take the job or not. Once that choice is behind us, our actions are largely determined by the requirements of the post. Typically, there is not much leeway left for personal discretion: what is to be done, when, and how are all prescribed externally. We thus feel pulled and shoved but inactive; it is difficult to think of ourselves *doing* anything without direct experience of the union of motive and execution.

An experience of impotence usually accompanies this sense of passivity. In mediated chains, the physical agent typically feels that the act is not his. He is, after all, caused to do it by his position in the chain. His own acts, in turn, are performed on his behalf by others far away. The longer the chain of intermediaries between himself and his distant act, the less he is in control of it. Even the most powerful man at the center of a large corporation time and again must taste this impotence. When things go awry, he suddenly comes face to face with his inability to control what his employees do. Often the most elaborate measures are inadequate to stop the slippage between command and execution. Even tightly organized social structures have an element of drift and, on the whole, the larger the structure the more difficult it is to regulate the behavior of its parts.

The third and perhaps most serious cost of mediation is what I call "psychic distance." As others perform my action, they and they alone come in contact with the circumstances that make it appropriate. Direct experience of the act resides in them alone; they see the immediate consequences, though sometimes even they do not see them all. The individual on whose behalf such actions are undertaken knows little about them; soon he may not even know that they occur. The result is a vast impoverishment of our personal experience and an immense, growing blindness to what we do.

Psychic distance makes us lose sight of the conditions of our existence and forget, if we ever knew, what many of our actions in fact and immediately are. Too frequently we do not know what we work, nor how it feels to cause what we condone. It is not that our minds have shrunk through willful anesthesia; instead, our bodies have grown larger or more numerous by the growth of society and by the cementing of those structures and relations which enable us to act on each others' behalf. The responsibility for an act may be passed on, but its experience cannot. The result is that there are many acts no one consciously appropriates. For the person on whose behalf they are done, they exist only verbally or in the imagination: he will not claim them as his own, since he never lived through them. The man who has actually done them, on the other hand, will always view them as someone else's and himself as but the blameless instrument of an alien will.

The psychic distance of which I speak is the direct result of the lack of direct experience. It shows itself in our unwillingness or even inability to appropriate actions that are clearly ours. It is reinforced by the fact that intermediate men hide from us both the immediate and the long-range consequences of our acts. Without firsthand acquaintance with his actions, even the best of men moves in a moral vacuum; the abstract recognition of evil is neither a reliable guide nor an adequate motive. If we keep in mind the psychic distance between the agent and his act, along with its source in impoverished personal experience, we shall not be surprised at the immense and largely unintentional cruelty of men of good will. The mindless indifference of what is sometimes called "the system" is in reality our indifference. It springs from our inability to appropriate acts as our own and thus assume responsibility for them—along with our bland perceptual life sheltered from encounter with evil. We do not know the suffering that is caused and cannot believe that *we* are the ones who cause it.

Our psychic distance from our deeds renders us ignorant of the conditions of our existence and the outcome of our acts. It fosters what seems to come naturally to most men anyway: blindness to the interconnection of all things, but especially of our acts and happiness. The distance we feel from our actions is proportionate to our ignorance of them; our ignorance, in turn, is largely a measure of the length of the chain of intermediaries between the original agent and his acts. The greater our ignorance, the more difficult it becomes to view the action as ours. And the less we are able to appropriate the act, the less inclined—in fact, the less able—we are to assume responsibility for it.

A tragic case demonstrating slippage of control, the danger of psychic distance, and the resultant problems of responsibility occurred during the Second World War in the Philippines when the Japanese General Tomoyuki Yamashita ordered his troops to do what was necessary to stop guerrilla activity.[1] It may be reasonable to believe, as he argued at his subsequent war crimes trial, that he did not mean for the troops to commit illegal atrocities and would have stopped them had he known what they were doing. The fact is that in spite of all the paraphernalia of modern communication, he was not in effective control of his troops, just as American commanders—with even better equipment—were apparently not in effective control at the time of the My Lai massacre. The loss of control is a direct consequence of mediation; its magnitude varies with the number of intermediaries involved (the chain of command) and the distance, both physical and psychological, at which policies ordered are executed.

The physical distance can, to some extent, be overcome by radio and telephone and television. The psychological distance is more difficult to handle, since it is itself a function of mediation and cannot be readily short-circuited. In order to minimize the likelihood of abuse, Yamashita should have known his officers. He should have had firsthand experience from the perspective of his foot soldiers of dealing with Philippine villagers and of moving for tense and frustrating days in a hostile jungle. Short of this, he could never have understood how an order that to him seemed professional and innocent could have been interpreted as a license for torture. The state of mind of the soldier on the line was infinitely distant from General Yamashita's intentions and understanding; without shared experience, the center cannot hold. With the best will all around, the person who executes the directive may become an executioner.

The War Crimes Tribunal understood all of this, but refused to accept the claim that as control slips and psychic distance increases, responsibility reaches the vanishing point. Yamashita was convicted and put to death on the assumption that as initiator of a chain of mediation, he was accountable for what happened all the way down the line. One can understand the sense of unfairness this verdict must have engendered in the general's heart. He must have known that he was being made an example of; persons lodged in the center of mediated chains elsewhere experienced the same slippage and caused deleterious consequences of the same sort, if not of the same magnitude, yet it would not occur to anyone to hold them responsible. The agency of one man simply does not extend that far, Yamashita must have felt. He was, of course, right in this. His agency did not extend so far as to be able to control his men. His intentions were honorable; he was a victim of circumstance. Yet the Court's verdict was strikingly wise in giving him not what he in fact deserved by current social norms, but what he ought to have got if there were justice in the world and had to get if there is ever to be a growth of responsibility in the teeth of demoralizing mediation.

Our sense of passivity in performing the actions of others, combined with ignorance of the nature and consequences of our own, make for what we think is a perfect excuse. We are not responsible, we think, because we live in the land between motive and execution, in an area of moral ambiguity. The Court's decree is notable because of its refusal to accept this moral namelessness. Both Yamashita and his soldiers had what might seem to the world a valid excuse. Each side could point to the other: the soldiers could claim that they were ordered to do what they did, while their general could self-righteously maintain that his intentions were innocent and trouble came only because they were misread. The situation was not unlike the one Steinbeck describes in *The Grapes of Wrath.* The men who come to evict farmers from their land seem to the victims as the enemy. Yet they claim innocence: they have been sent by the bank, they are only doing their job. If they did not act faithfully on behalf of their employers or if they were shot, others—perhaps far worse than themselves—would be hired to take their place. The local banker himself is not to blame. He has his bills to pay and takes *his* orders from others far away. And those far away are, presumably, in the same bind. No one means ill, everyone is caught; we are all innocent parties to an agentless crime. The desperate farmers go straight to the heart of the matter. If all of this is so, they ask, then whom do we shoot?

The War Crimes Tribunal's answer to this is that you shoot the man who is at the relative center of the chain of mediation. To do so is to reject the claim that psychic distance and the loss of control entitle us to reject responsibility. From the standpoint of the classic maxim of moral philosophy that one is obliged to do only what one can, Yamashita was blameless. Given his circumstances, the social practices in the midst of which he found himself, his education and his past experience, it was simply not within his power to control what went on in the field. The Court's verdict is so striking because it simply disregards this very real inability: it says, in effect, that he should have controlled his troops, whether he was able to or not. He should have put himself in a position where he could control them, or else suffer for their misdeeds as though they were his own. The message to all of us is that acts done at our behest, no matter at what distance, are ours, that mediation must not be allowed to destroy the gossamer fabric of responsibility. The Court was not convicting upon the basis of existing norms of morality; it boldly engaged in the enterprise of creating new and more stringent standards.

Evidently, there are very few instances of irresponsibility due to psychic distance in corporate life which are this dramatic. Yet the same concepts which render the Yamashita case intelligible can be used to shed light on the causes and cures of employee irresponsibility. The mediated chains of which corporations consist naturally create a momentous psychic distance between individuals in the chains and those acts which are performed on their behalf and without which the corporation could not survive. In ordinary terms, this means simply that just about everybody in a large company is ignorant of the detailed interconnectedness of its operations. The nescience is particularly pronounced among those who serve on a relatively low level: most of the decision making, integration, and social impact of the company are typically beyond their horizon. But there is a corresponding ignorance among the decision makers. These may well understand the outlines or principles of the manner in which the corporation produces and markets goods. Yet they have little knowledge of the actual experience of production. They have no just sense of the toll menial jobs exact or of the debilitating effects of routine. The daily boredom of managers escapes them, as does the way in which narrow role and stressful home combine to create lives without content and prospect.

The acts of all these agents constitute a single whole: they are so interconnected that it is not inappropriate to read each as being performed on behalf of many or all the people in the company. Yet this interconnectedness is not understood. The worker does not grasp the rationale behind management decisions and thus views them as alien; management, in turn, sees worker demands as self-serving and unacceptable. The result is the same: each person in the chain finds the perfect excuse by disappearing in his job. Each can convince himself that he is a victim of circumstance, an insignificant cog in the machine. How can anyone be held responsible if he is not an autonomous source of activity, if agency—so to speak—merely flows through him the way it flows through each link in a steel chain?

Responsibility is inconsistent with anonymity. Yet links in a chain of mediation have no names and no personality. All that is caught of the man is the role he plays, the job he performs: he is salesman, mail clerk, concentration camp guard, or sergeant assigned to command an execution squad. And paradoxically, the more we fade into our jobs and disappear in them, the less we identify with them, the less they express us, the less they satisfy. If my job is to serve as a shield from moral accountability, I must be utterly uninvolved with it: it must always appear that I am a cool, impersonal agent intent on doing what I must. In this way, job activity is mated to personal passivity, and our institutional obligations become the blanket excuse we offer to those who would hold us humanly and morally responsible. "It is nothing personal, you know," one can hear the Lord High Executioner whisper to his victim or the manager to the employees he is about to fire. "It is just my job to do this," we all say. "Who am I to question the wisdom of it all?" To the skeptical, we might add, "I owe an obligation to the company that pays my way," and then hasten to the ultimate disclaimer, "and in any case, if I did not do it, they would simply get someone else."

The high note of obligation to the company is, of course, a sham. For most, the duty extends no further than the narrowest limits of the job. Irresponsibility beyond the corporation and irresponsibility within it are of the same fabric. The same absorption of the man into his job that protects from responsibility to customers and the society beyond also protects from any but the narrowest accountability within the corporate structure. In this way, "I only work here" is in natural alliance with "that isn't my job." Both forms of irresponsibility have their origin in the psychic distance that is a natural outcome of mediation. Both suggest an inability to appropriate actions performed on our behalf, an unawareness of the nature and consequences of our acts, a very real ignorance of the interconnectedness of our lives.

If this analysis of the causes of employee irresponsibility is accurate, the cure should not be difficult to find. It is obvious that mediation cannot be abandoned. To hanker for smallness is to express a romantic dream. In any case, to give up the large-scale mediated structures of our industrial society would entail a cost so staggering that it outstrips the ability of our imagination to measure it. The only hope is to reduce the psychic distance that attends mediation. In an unconscious way, some of this is already being accomplished for potential managers. Training programs in banks and other institutions take young people through the entire hierarchy from the mail room to the board room. The trainees learn to do each job and actually do many of them for a stretch of time. The avowed purpose of doing this is laudable: someone who will soon be in a management position has every reason to learn the way in which his corporation operates. The important thing to realize is that such exposure yields not only factual knowledge. A possibly unintended benefit is that by understanding the nature of each operation and the interconnectedness of them all, the manager begins to think of the corporation as *his*.

By saying that he views the entire company as his, I do not mean he supposes that he owns it. The true meaning of ownership is not the formal one we have enshrined in the law. It is something far more personal and immediate, which in the social or corporate context involves at least three abilities. The first is to see the complex operations of others as conditions of our own activity; the second is to think of their work as beneficial to us and as being done on our behalf; the third is to view the whole that their work and ours compose, as an expression of our creative life.

It is this vision of the whole and a just appraisal of our part in it that we must promote, if irresponsibility is to be overcome. The mail clerk needs this no less than the president. It is not that by understanding how he contributes, the mail clerk will think his job more lofty or exalted. It is just that he may begin to take due pride in the fact that it is as indispensable as the decision-making of the chairman of the board. He may also develop a better grasp of how the consequences of his acts may hinder or endanger the entire enterprise. And if he understands that, it may suddenly dawn on him that in hurting the whole he is—in however roundabout a fashion, yet very really—hurting himself.

Mere understanding is not enough. Lectures on how the corporation works never suffice to overcome psychic distance. Immediate exposure to what others do and how is essential in rendering our abstract knowledge vivid and effective. But in addition to this increase in immediacy, we need also to give employees the genuine sense that they are participants, nay partners, in a joint enterprise. This requires the introduction of an element of grassroots democracy into the corporation. For employees to feel that the corporation is somehow *theirs* (even though legally they do not own it), they must sense that what it does is to some little extent the result of their decisions. Obviously, whenever large numbers are involved, the influence of any one individual must remain minuscule. Yet the very fact that on some important issues each employee is offered a say increases the likelihood that he will be able to appropriate the course of action that emerges as his own. People do not want to be told the lie that their decision in the corporation or in the state can be decisive. They know that their view is but one of many, and what frustrates them is not that what they want fails to come about but that they rarely get a serious hearing. It is the serious hearing that should always be provided, along with realistic information about the presence and strength of many conflicting views.

One might think that such autonomy within business organizations would be exceptionally costly. This is both true and false. Admittedly, time would have to be devoted to educating employees and to canvassing their opinions. But whatever loss this might involve would be more than made up for by increased readiness to assume responsibility, increased enthusiasm, and increased productivity. This is the lesson of the failure of collective farms in Eastern Europe. Their productivity proved to be a disaster compared to what happened on the small plots the state left to individuals and families. In their home gardens, the peasants worked efficiently and well; they saw what they did as somehow their very own. By contrast, they could never view the collective farms as extensions of their personalities, as in any important sense belonging to them. The result was the inefficiency that comes from not caring, the surly negligence typical of those who are made to do things and thus feel put upon.

We can avoid this fate and improve productivity by making corporations psychologically the property of all who work in them. Making them shareholders will not necessarily give workers the sense that they have a stake in its welfare. Teaching them how the corporation works and asking them to participate in running it frequently will. The choice is between full participation and continued growing irresponsibility.

NOTE

1. The two most instructive documents on the Yamashita case are A. Frank Reel, *The Case of General Yamashita* (New York: Octagon Books, 1971), and Courtney Whitney, *The Case of General Yamashita: A Memorandum* (Supreme Commander for the Allied Powers. Government Section, 1949).

13

CONFLICT OF INTEREST:

Whose Interest? And What Conflict?

JOSEPH M. McGUIRE

The disclosure of conflict of interest cases in almost every major American institution during the past decade has proved to be distressingly newsworthy. Next to mugging in the urban core and burglary in our pastoral suburbs, such activities as payoffs, kickbacks, bribes, "contributions," graft, and similar gentlemanly but deceitful practices by "respectable" members of society would appear to be among our more popular criminal pastimes. Conflicts of interest have created difficulties in government, business, most of the professions, and even in higher education. Disclosures of these conflicts have toppled highly placed public figures in the United States, in the Netherlands, and in Japan. Chief executives in giant corporations have been forced out of their prestigious posts in disgrace. Public accounting firms, which in recent years have increasingly appeared in the courtroom, often have alleged roles in conflict of interest situations. Doctors of medicine, in this day of health insurance and Medicare, have frequently been accused of placing their financial interests above their patients' interests. The conflicts of interest in our religious institutions are most obviously seen in the statistics on priests and nuns who have forsaken their vows for a more secular life style. As only one illustration of the many conflicts of interest in higher education: Professors who force their captive classes to purchase the texts they have authored have increasingly become subject to the scrutiny of students and university authorities alike. No institutional setting today, it would appear, is free from the taint of well-publicized conflicts of interest.

Scandals associated with conflict of interest disclosures have, in turn, led to a huge ongoing reform movement—a sort of ethical housecleaning—resulting both in legislation and in the writing or rewriting of codes of ethics for public agencies, corporations, and some of the professions. In the public arena, for example, forty-four states have passed new political reform laws since 1972.[1] Campaign financing has been more tightly controlled, with cumbersome legislation designed to prevent the domination of candidates by wealthy vested interests. Corporation officers and directors are strictly prohibited from private gains garnered through inside knowledge. Lobbying activities in many states have been severely curtailed. The *Wall Street Journal* reports that "Accountants Reassess Disclosure Standards After Business Scandals."[2] It appears, therefore, that we are now at the crest of the wave of reform directed toward the avoidance of potential conflicts of interest and the punishment of those individuals or organizations that yield to their prurient instincts in such situations. As Richard L. McAnaw recently observed, however: "Reform, like pestilence, arrives with each generation. Not infrequently it arrives at the wrong time, for the wrong reasons, and leaves behind the wrong remedies."[3] This may well be the case with the reforms directed at conflicts of interest, as will be noted later in this essay. First, however, we shall explore the concept of conflict of interest itself. And, as the title suggests, we shall find eventually that it does, indeed, make a great deal of difference as to what the conflict is about and whose interests are involved.

THE CONCEPT OF CONFLICT OF INTEREST

The term "conflict of interest" has become widely used in recent years, and is probably understood by most members of the adult population in the United States much in the way they understand such terms as "love," "faith," or "electricity." To define it precisely, however, so that it can be probed and analyzed is another matter. In fact, this situation is somewhat reminiscent of the story told about James Thurber, who, when informed by a lady at a cocktail party that she much preferred to read his books in the French editions, is reported to have murmured: "Well, they *do* lose something in the original." In the same way, it might be better to leave the meaning of conflict of interest in its public translation obscure and vague . . . but generally understood. Unfortunately, I am compelled by my many years in academe to be more precise, but something undoubtedly will be lost in this original too.

Conflicts of interest are imbedded in human nature, and illustrations of them abound in the history of our own nation. For example, George

Washington Plunkitt was a famous Tammany Hall ward boss renowned for his very frank talks on very practical politics around the turn of the century. One of his most quoted speeches was entitled "Honest Graft and Dishonest Graft." In it, he differentiated between the two. Dishonest graft, he said, was practiced by "blackmailing gamblers, saloonkeepers, and disorderly people." Plunkitt, on the other hand, righteously observed that he practiced only honest graft—that which resulted from the judicious and shrewd use of inside information, primarily about the land acquisition plans of New York City and its agencies. Before his death, this millionaire octogenarian remarked that he wanted for his epitaph: "George W. Plunkitt: He Seen His Opportunities, and He Took 'Em."[4]

A conflict of interest, in the public mind, may be synonymous with honest graft. However, to cover a wider range and variety of cases than those resulting only from the possession and use of inside information, we shall express our definition in the jargon of systems theory. Thus, *a conflict of interest exists when a subsystem attempts deliberately to enhance its own interests or those of an alien system to the detriment of the larger system of which it is a part.* The use of the word "subsystem" broadens and depersonalizes the definition, but in no way is it intended that the individual be excluded. A subsystem, therefore, may be a person, but it also may be an agency of government, a division, department, or office in a corporation, or the member or group of any organization or society. An alien system is typically another organization, but it may be any person, nation, or other unit that is not an integral part of the larger system. Furthermore, this definition implies not only that the values and goals of the system and the subsystem are different, but that they must be diametrically opposed rather than merely complementary, so that attempts to increase subsystem gains will harm the interests of the larger system.

As the definition now stands, unfortunately, it does an injustice to what is commonly understood to be a conflict of interest, largely because it does not *a priori* judge that the interests of the system are "good" and "desirable," and the interests of the subsystem are "bad." In other words, the definition is flawed because it is value free. As such, it covers the generally accepted meaning of conflicts of interest. For example, the situation wherein the attempts of an alcoholic member to increase the duration or extent of his or her euphoric stupor harms the family system is included under this definition. So too is G. W. Plunkitt, seizing opportunities to purchase land at low prices and to resell it at high prices to the detriment of the community, covered. And those cases wherein directors profit financially from inside information at the expense of investor confidence and the corporate good. It is probable that, in these and similar cases, there would be general public agreement that conflicts of interest exist. However, it would not be correct to jump to the conclusion that, because these illustrations conform to the public impression of conflicts of interest, the definition above is satisfactory.

The problem with our definition is that it includes too many types of conflict which do not conform with the public image. For example, it covers those situations which might be considered ethically "neutral," such as the conflicts that the organizational models of Robert K. Merton and Philip Selznick consider normal.[5] Both of these organizational models take for granted the appearance of dysfunctional attitudes whereby subsystems establish barriers to the attainment of system goals. In the Selznick model, for example, the individuals in large and complex organizations will "naturally" resist efforts by top management to use them as means to organizational goals, and will attempt to substitute their own purposes for those of the larger system. In the Merton and Selznick theories, the attempt to advance the purposes of the organization results in unintended and undesired individual and subsystem behavior. Such conflicts, however, would probably not be included in the public notion of conflict of interest, although they are covered by our definition.

Most importantly, our definition of conflicts of interest includes those cases where right is on the side of the subsystem. This category of conflict situations would not ordinarily be included in the public definition. The individuals described in John F. Kennedy's book, *Profiles in Courage,* were elements of subsystems eventually applauded for their brave words and actions in the light of system goals. As the Nuremburg trials attest, individuals have been castigated for their failure to enter into conflicts with the larger system. The American public typically would approve the attempts of a small band of democratically inclined revolutionaries to overthrow a dictatorship.

The point of all this is, very simply, that: while there is nothing inherently wrong with our definition of conflict of interest, it does not conform with the commonly accepted notion of this term because it does not contain *a priori* ethical assumptions about the "rightness" of system and subsystem interests, goals, and values. In order to make our definition correspond to the public concept of conflict of interest, it would be essential to assume that the interests and values and goals of the system are ethically good and morally right, and that those of the subsystem are ethically wrong.

And, indeed, it is this underlying presumption of the ethical propriety of system goals, interests, and values that makes the concept of conflict of interest so complex and bothersome. It is this presumption, too, that may cause reforms to be undertaken ". . . for the wrong reasons and [to leave] behind the wrong remedies." Our definition of conflict of interest, if ethics must underlie it, is situational in nature. It does not prejudge whether the subsystem or system is right or wrong. In some situations, the subsystem conflict activities may be morally correct; in others, immoral. In the public concept, however, right always is attributed to the system. The latter rests upon an absolutist, or sometimes cultural, concept of ethics. Thus, espionage activities are basically conflicts of interest. However, when American spies are used in wartime situations, the public applauds them as heroes, while alien spies in the United States are obviously in the wrong.

Whose interests are harmed by conflicts of interest? The interests of the system. Is this good or bad? According to our definition of the concept, the results of a conflict of interest may be good, bad, or ethically neutral. To the public mind, however, it would appear that harm to the system is always wrong.

Given the public attitudes toward conflict of interest situations, therefore, it is not surprising that the remedies for rectifying those situations have consisted largely of removing temptation from, and imposing sanctions on, subsystems. We shall examine some of these reforms in the following section of this essay. At the same time, however, it is necessary to note that such reforms are not costless. There might be less freedom of speech and action in organizations and more demands for conformity to organizational values and goals. There will be a greater hesitancy on the part of some good people to take certain positions because of the "appearance" of conflicts of interest. There may be the imposition of righteous and rigid moral codes upon organizational members. And, finally, many of the efforts at reform are based on a rather dismal view of man, regarding people as basically weak, not to be trusted, and with a natural propensity to place their personal interests above the system's "good." We shall pursue these "costs" of conflict of interest reforms as we examine the reforms themselves in the following section.

CONFLICT OF INTEREST REFORMS

Recent attempts to prevent conflicts of interest or to punish participants in them appear to fall into one or more of the following categories: (1) efforts to reduce the potential for conflicts of interest; (2) attempts to write or rewrite codes, guidelines, and policies to make clear and explicit what is not permitted by the organization so that members will know what the rules are and when their activities will be in conflict with these rules; and (3) legislation. We shall examine, through illustrations, each of these classes of reform.

1. Reduction in the Potential for Conflict of Interest

The furor over conflicts of interest has led some individuals and organizations to adopt an "arms-length" policy whereby they attempt to avoid all situations in which there exists the potential for conflict. This attitude has apparently produced a relatively widespread "don't question company goals" sentiment among corporate managers. For example, Archie Carroll recently conducted a survey of managers at various levels in approximately 240 randomly selected companies. One of his findings was that almost 65 percent of these executives believed they were currently under considerable pressure to compromise their personal standards to achieve company goals. About 60 percent of the respondents agreed that young executives in business would have loyally gone along with their bosses, just as the junior members of President Nixon's reelection committee did.[6] These remarkable results attest to the opinion, earlier expressed, that one way to avoid conflicts of interest is to sublimate one's own interests to those of the organization. If one is blindly unswerving in his or her loyalty to the organization, the only conflict that might occur is the inner one—but over time, probably even this trauma can be lessened if only one perseveres.

More sensitive individuals and organizations evidently are reducing the potential for conflicts of interest by simply refusing to serve or to ask others to serve. Thus, in a recent survey by Korn/Ferry International (an executive recruiting company) of 407 major corporations, it was found that the number of lawyers, commercial bankers, and investment bankers on the boards of directors of these enterprises had declined between 1973 and 1975. *Business Week* reports John L. Hanigan, chairman of Brunswick Corporation, as saying:

> I don't think we should have partners of investment banking firms or executives of commercial banks on the board. You have to ask, "Whose side are they on" They sure have a conflict because they are trying to sell their services to you.[7]

A survey by Booz, Allen and Hamilton, Inc., of 30 top accounting, consulting, investment banking, and law firms found that only 6 of these companies encouraged their members to accept directorships, while 14 discouraged acceptance and 11 were neutral. Clifford E. Graese, a partner in Peat, Marwick, Mitchell and Company, says:

> We want to avoid even the appearance of conflict of interest. The potential for conflict is too great. We could not have one partner on a bank's board, while the firm audits another bank.[8]

The sensitivity of these groups and individuals to serve on corporate boards has not been without its side effects. Thus, academicians, women, members of ethnic minorities, economists, and representatives of similar groups have increasingly become members of boards of directors in place of the lawyers, bankers, accountants and others who have refused, or been refused, membership.[9] Presumably, these latecomers are not concerned about the possibility of interest conflicts, or possibly none exist.

Some observers of these developments believe that the loss of legal, financial, and professional talent on corporate boards will make these boards (and, presumably, the companies themselves) less efficient. David C. David, former dean of the Graduate School of Business Administration at Harvard University, is reported to have said, for example, "that in his experience the men with the conflicts usually make excellent directors."[10]

Even cynics like Edward S. Herman, professor at the Wharton School of Finance, feel conflict of interest radicals are shooting at a straw man. "They confuse form and substance," he says. Most of those outsiders are there for tacitly understood, limited functions—which don't happen to include control."[11] And finally, John M. Schiff, general partner of Kuhn, Loeb, and Company, remarks:

> The more they get their computers cranked up in Washington, the more conflicts of interest they will find. But whether they are real conflicts or not remains to be seen. I think the attack is based on a philosophy that every man is basically dishonest, and that I don't believe.[12]

While the members of certain professional and management groups have become increasingly timid about accepting corporate directorships where the potential for a conflict of interest exists, evidently this same apprehension has not spread widely into other areas of organizational activity where joint or sequential affiliations have the potential for conflict. One of the most tenacious of these situations is found in the movements of individuals between federal regulatory agencies and their regulated industries. A recent study by Common Cause, for example, reports that of the 36 commissioners who left regulatory agencies in the last few years, 17 returned to companies their agencies had regulated or to the companies' law firms. The same study revealed that 279 of the 429 top employees at the federal Nuclear Regulatory Commission came from companies which deal with the agency. And, finally, Common Cause found 518 employees in 11 agencies who had financial interests which appeared to conflict with their duties, despite the existence of governmental disclosure rules.[13] A House Commerce Subcommittee found that in the period 1971–1975, 22 of 42 nominees for regulatory agencies had been employed directly or indirectly in industries regulated by the agencies for which they were nominated.[14] Subcommittee Chairman John Moss said:

> While in any individual case, a commissioner may perform without bias . . . the practice of selecting so frequently from the portion of the private sector to be regulated cannot fail to erode the credibility of a commission with the public.[15]

Although the evidence suggests that potential conflict of interest on corporate boards is increasingly recognized by the participants themselves,[16] the potential conflicts arising from the shuffle of executives back and forth between regulatory agencies and regulated companies—movements which have been criticized by academicians and others for decades—appear to continue unfettered by the shackles of legislation, public opinion, or private morality.

Common Cause, as well as other public interest groups, has proposed that more stringent financial disclosure rules for government regulatory agencies be written and enforced, and that the White House broaden its recruiting search for new talent. Oddly enough, however, there seems to be more tolerance of potential conflicts of interest in this area than in the purely private sector, possibly because, as Common Cause itself has noted, the regulatory agencies do indeed desperately need the expertise of people who have had experience in the regulated industries.

The United States is not the only nation confronted with potential conflicts of interest on the part of government officials whose backgrounds or investments may bias their decisions and actions. In 1975, for example, the House of Commons agreed that all members of the British Parliament should file a statement indicating the sources (but not the amounts) of their personal incomes or other interests, to be published at least twice a year. As the *Economist* asked:

Are we imposing on politicians an unjustifiable invasion of privacy? Do we want MPs to have no other financial interests, and to be the sort of people who invest only in savings bonds? Can private conflicts of interest, consciously or unconsciously, sway the policies of British governments?[17]

And as the article concludes:

The . . . question is whether one believes with Plato that incorrupt government comes best from politicians who have no income save that from the state, or Lord Curzon's aristocratic dictum that no man should be viceroy of India to whom that job is an honour. The answer is that, except for a few hereditary millionaires, British government has depended since the war on attracting men who would gain a decent bourgeois standard of life by their own efforts if in outside occupations; and the blunt convention has been that many politicians do attain this by business or other emoluments that are given to them. . . .[18]

Efforts to solve potential conflict of interest situations typically consist of proposals to prohibit members or former members of alien systems from membership in "the system." The basis for such denials is that the possible allegiances to the alien system will be, or might be, troublesome. In dealing with conflicts of interest, then, as in other areas such as liquor, drugs, and pornography, mankind stands ready to protect the individual from his baser cravings. And, again, society stands alert; prepared to re-create the Garden of Eden by removing, in potential conflict cases, the possibility of temptation. This Victorian mentality regards mankind as weak and seeks to minimize the exposure of all individuals to the occasion of sin. As in similar situations, the blanket and heroic generalizations about the nature of mankind's frailties preclude the best and strongest, as well as the weak and unethical, from accepting positions of responsibility. Such assumptions, at their best, inhibit freedom of choice; differentiate between individuals on the basis of what may, indeed, be irrelevant but all-important criteria; impugn the image of man as an individual and ascribe to each person an all-important social identity; and find each person guilty of weaknesses in the face of temptation. It is difficult to understand why the efforts to protect persons from potential conflicts of interest should be condoned. Whether an individual will succumb to temptation in these situations, it would appear, depends more upon the character and integrity of the person than upon the nature of the conflict. To solve potential conflicts of interest through universal prohibition would appear, therefore, a somewhat childish way to prevent such situations.[19] We shall not stoop to discuss the prevention of potential conflicts of interest through the sublimation of personal integrity to organizational interests, since the evil of such a neat solution to the problem must be evident to all.

2. Codes, Guidelines, and Organizational Policies to Prohibit Conflict of Interests

Many years ago, Commerce Secretary Luther Hodges, when pressed for the date of release of a departmental code of ethics for American business, said: "It's hard to write a new ten commandments."[20] And, as I noted a year or two later:

> Even the most ardent Democrat has probably never regarded Secretary Hodges as a second Moses, and it is doubtful that anyone ever would expect the lightning fingers of the Department of Commerce, fiery though these may be, to emulate successfully the code of a higher but less bureaucratic Being.[21]

Nevertheless, the statement by Hodges does indicate the difficulties of writing prescriptions for corporate behavior—a task which, if I recall correctly, was never completed. The very fact, however, that the Department of Commerce found it necessary to add still another code to the list attests to the improbability of ever achieving any creed which will be wholly satisfactory.

In this part of this essay, however, we are not going to explore the task of writing an epistle to cover globally all possible areas where transgressions might occur in organizational life. But, in a somewhat limited way, we shall raise the question: "Is it possible to write (or rewrite) codes that will prevent conflicts of interest? And can these codes be effective?

Raymond Baumhart, S.J., in conducting his well-known study of business ethics in 1961, included in his mail questionnaire to 5,000 executives several inquiries on ethical codes.[22] For example, Baumhart found that formal company policy was next to a man's personal code of behavior as the most important ethical influence on his actions and decisions. A lack of company policy, on the other hand, was among the least important influences causing executives to make unethical decisions. Over 70 percent of Baumhart's respondents believed that codes of ethical practices in each industry would be a good idea. However, while such codes would raise the ethical level of industry, help executives by defining clearly the limits of acceptable conduct, and be useful aids when executives wanted to refuse unethical requests, most respondents also believed that such codes would be extremely difficult to enforce, and that people would violate them whenever they thought they could avoid detection. Surprisingly, perhaps, the majority of respondents also believed that these codes would operate more effectively if they were enforced by non-government regulatory groups external to the company rather than by company management. The cynicism and distrust evident in the responses about enforceability and the preference for regulation by external groups would appear to make even the least skeptical of Baumhart's readers doubtful of the utility of ethical codes, although it would indeed appear more satisfactory (for public relations reasons, if for no other) to have them than not.

Some conflict of interest situations, on the other hand, can be satisfactorily covered by organizational policies and guidelines. In two cases in particular—the extent of external affiliations and the acceptance of gifts by organizational members—organizational policies are frequently effective. Consequently, many businesses and government agencies have, at the very least, disclosure policies, wherein members are required to inform management or to make public their financial interests in other organizations. Some companies explicitly forbid their employees to have investments in suppliers, customers, or distributors. Others permit employees to own some percentage of stock in such organizations. Still others require disclosure of all outside business interests—financial, kinship, or whatever—by key executives and purchasing agents. Organizations often are fearful, too, that the acceptance by their members of gifts or entertainment might lead to a conflict of interest whereby the recipient might be disposed to give preferential treatment to the giver, and therefore frequently have policies covering these situations. In some organizations, employees cannot accept any gifts from external sources. In others, the value of the gift, its purpose, the circumstances surrounding it, the position of the recipient in the organization, and general practices in the organizational culture determine the policy that is established. Thus, in one company it might be acceptable to receive a calendar or a pencil with the donor's name on it, while in another there may be specific prohibitions against even such nominal gifts. Many organizations set an upper dollar value on gifts that might be accepted, assuming, implicitly, that items costing less than $10 or whatever will not result in preferential treatment for the donor. Entertainment of organizational members by alien systems also is frequently controlled by guidelines. Organizational rules in this area also vary from a strict prohibition against the acceptance of payment by others of the member's food, drink, etc. to relatively lax rules which permit Congressman Ford to enjoy golfing holidays.

Policies and guidelines designed to control situations which might create interest conflicts would appear to be desirable for all organizations, but it is clear that the totality of such conflicts is difficult to circumvent by organizational codes alone. Typically, if such codes are too strict, they inhibit member effectiveness, for they interfere with social conventions. On the other hand, if they permit reasonable latitude in member behavior, they are likely to be misinterpreted by the best of men and deliberately broken by the worst. In the last analysis, such codes, since they attempt to ward off temptation, have the same shortcomings as the efforts to remove the potential for interest conflicts discussed earlier. The one advantage that such codes have, however, is that they do provide limits beyond which organizational sanctions may be imposed, and thereby tend to act as a deterrent to some of the weaker organizational members.

No code or policy or guideline, however, can be expected to cover all conflict of interest situations, or to prescribe behavior for the great variety of such situations that might occur. In fact, there are many such cases where codes would be undesirable. In higher education, for example, publishers have long given conference cocktail parties and "desk" copies of texts to faculty members. In my years in academe, I have never met a professor who refused such favors because he or she feared that acceptance would lead to a conflict of interest. At the same time, I am not so naïve as to believe that publishers proffer such gifts and hospitalities out of their love and admiration for members of the teaching profession—a group which even I, as a member, have often found difficult to tolerate over the years. Nor, to give one last illustration from higher education, is it possible for policy to cover what I term the "administrative sickness" of many deans, chancellors, provosts, and presidents in our universities, wherein they consciously or unconsciously are involved in the conflict between self-aggrandizement and the duties of their positions. To perform the latter properly, they should spend most of their time on campus or with the university family (e.g., legislators, trustees, alumni, community leaders, faculty), but instead, they are often drawn by their dreams to address the Economics Club of Punxataway, Pennsylvania; to participate in a panel of the ACE on the quality of student food; to serve as an advisor to the National Science Foundation; or to travel to Japan for the Ford Foundation. The best administrators put their tasks first, of course, although there are some, also good, who simply enjoy these perquisites. Unfortunately, most administrators who do all these things argue that there is no conflict at all, although some of them must feel guilty.

Should there be organizational policies governing situations that might lead to conflict of interest? By all means. Will these be effective? Possibly, but we should not hope for too much. Is it possible for such policies to cover all conflict situations? The answer is negative. Let us, by all means, have such organizational codes. They are helpful. But let us not expect too much from them.

3. Legislation

In the mid-nineteenth century, Cornelius Vanderbilt petulantly asked: "You don't suppose you can run a railway in accordance with the statutes, do you?" In the seventh decade of the twentieth century, the question would seem to be whether any organization, including the government itself, can operate in accord with the statutes. Certainly the evidence, especially that surrounding such issues as conflict of interest, would appear to indicate that good legislation is as difficult to write as a good man is to find, and that organizational conformity to such legislation is almost impossible to enforce unambiguously and with equity and justice. Nevertheless, new laws have been forthcoming in recent years to prevent conflict of interest situations and to impose penalties upon transgressors, and it is likely that legislation will continue to be written in the future.

Reform through legislation had been made inevitable by Watergate and by the disclosure of payments to foreign agents and domestic politicians by American corporations. Such payments, of course, have been customary over the years, but their magnitude, as disclosed in the mid-1970's, was shocking to the public and to government officials alike. For example, the Chairman of the Securities and Exchange Commission, Ray Garrett, was quoted as saying:

> . . . bribery, influence-peddling and corruption on a scale I had never dreamed existed. One suspected the standard things—the padding of expense accounts, with the understanding that the excess was to go for political contribution. The payment of minor "grease" as it is known in my home city (Chicago). The wining and dining. But none of us dreamed there were the millions, the tens of millions, the hundreds of millions, we have found.[23]

Garrett's astonishment followed the disclosure of a number of cases wherein major corporations had paid huge sums to foreign agents in return for promised benefits. Among these cases was one involving the Lockheed Aircraft Corporation, which paid $202 million to political parties, officials of foreign governments, and sales representatives.

Senator Frank Church speculated that $30 million of this was in the form of "pure" bribes. Among smaller, but still sizable, similar payments were the $450,000 bribes of two Saudi Arabian generals by the Northrup Corporation, the $1.3 million paid by United Brands to Honduran officials, and the gift of $50,000 by Ashland Oil, Inc., to three representatives of the Dominican Republic.

Much has been written and said about these and similar instances of bribery and corruption in recent years. The rhetoric—to say the least—has been indignant. The legislation, however, to rectify these types of problems has been slow to materialize on the federal level, and the laws that have been passed seem to constitute relatively minor reforms. The SEC, for example, has put increased pressure on corporations and their outside auditors to disclose publicly all contributions abroad, for the most part relying on the concept of materiality, which requires disclosure of "any fact which a reasonable investor might consider important in making an investment decision."[24] The SEC has also used injunctions whereby executives, while not required to acknowledge past sins, are made to promise not to commit them again in the future, and some recent laws provide for fines and jail terms for executives who violate them. For the most part, however, federal legislation on conflicts of interest in recent years has not been overly interesting, impressive, or innovative. Thus far, the federal government has relied largely on old legislation and hasn't been able to pass new "young" laws. As Thomas Griffith has observed:

> Whether the area of corruption is in defense, oil, or merely bananas, the continuing bad publicity, and the feeling that crime in the executive suite is gently punished, greatly increases the present-day cynicism about both business and politics.[25]

Much of the federal legislation that has been passed with regard to conflicts of interest has been directed toward the government system itself, or to its interface with external institutions, rather than toward external institutions such as businesses. The states, however, have been more active than the federal government, and in many ways some of their legislation has been more imaginative and interesting. We shall examine some of these laws, drawing our illustrations from the State of California.

The California Political Reform Act of 1974 requires that

> No public official . . . shall make, participate in making or in any way attempt to use his official position to influence a governmental decision in which he knows or has reason to know he has a financial interest.[26]

"Financial interests" are defined as direct or indirect investments of $1,000 or more, income totaling $250 or more received during the preceding year, or any company of which the public official is a member. The Act requires public officials to file statements of their financial interests annually with the Fair Political Practices Commission, an agency established by the law. The Commission is authorized to implement the Act, hold hearings, and, if it finds a violation has occurred, to issue cease and desist orders and/or impose fines up to $2,000.

Another part of the Political Reform Act of California requires campaign finance disclosure reports of the employer and the occupation of every contributor of $50 or more. Auditing and enforcement provisions can result in penalties against businesses that range up to $10,000 fines.

Perhaps the most discussed section of the Act is that which prohibits lobbyists from making gifts of more than $10 per month to state legislators or administrative officials. They are also required to register and to file complex monthly reports of their activities. Furthermore, under the provisions of the Act, any person who attempts to promote, support, influence, modify, oppose, or delay state legislation or administrative action is lobbying, and as soon as he or she spends $250 on such activities in a month (including the purchase of newspaper advertisements) a lobbying report must be filed.

Despite the fact that the Act contains over 20,000 words of definition, description, and discussion, there is considerable uncertainty still about the meaning of the law. For example, soon after the passage of the Act, the Fair Political Practices Commission spent a month just trying to define what a private-interest lobbyist is. These uncertainties have caused many companies and associations to take great precautions to conform with the law. Now, for example, the Pacific Gas and Electric Company has registered seventeen lobbyists, as compared with five in 1974.[27] The California Taxpayers Association, unable to differentiate between research personnel and lobbyists, has registered all professionals in its Sacramento office. A number of companies initially failed to register at all as lobbyists, but have now done so, although they remain uncertain of their status. The Commission has been sued by such disparate groups as the California Bankers Association and the Socialist Workers Party. One of the more serious questions is whether lobbyists, such as those employed by the California Labor Federation, can advise their membership about legislators' voting records without breaking the law.

Some groups and individuals believe that the California Act, especially its lobbyists' provisions, interferes with that phrase in the First Amendment which guarantees "the right of the people . . . to petition the government for a redress of grievances," and that disclosure requirements imposed upon private persons and organizations violate the Fourth Amendment's privacy provisions.

Earlier comments about potential conflicts of interest and organizational policies to prevent conflicts or punish transgressions apply to much of the legislation that has been enacted at the state and federal levels in recent years. This legislation typically presumes not only that elected or appointed officials or bureaucrats must be "pure" but that they must prove their purity. It assumes they cannot withstand the blandishments and temptations and bribes of representatives of external systems, so the latter must not be permitted to offer the former apples—or anything more expensive. In order to keep these tempters away, legislation is devised that impinges on freedoms guaranteed by the Bill of Rights. It is doubtful that legislation will deter either legislators or other transgressors who are determined to corrupt or be corrupted. It may, however, prevent more ethical individuals from marginal, controversial, or even healthy activities. It is also doubtful that legislation will cleanse our political system of conflicts of interest. Persons who are capable of operating "off the record," and who are disposed toward unethical behavior, will undoubtedly continue to do so. Legislation will, however, make those who are crafty and unethical more careful.

CONCLUSIONS

If the remedies we have discussed, designed to resolve conflicts of interest, are not particularly attractive, why is this the case, and what might be more desirable solutions?

First, it should be recognized that not all conflicts of interest are undesirable. The definition presented in the first section of this essay, and the discussion surrounding it, tried to make clear and precise that attempts at subsystem gains are not necessarily unethical or evil. But, underlying the public concept of this term, is the presumption that such subsystem activities are per se "wrong." In their efforts to correct conflict of interest situations, therefore, private and public policy makers have attempted to prevent all conflicts of interest—the meritorious along with the harmful, the innocent or innocuous or "natural" along with the guilty. History has shown the folly of condemning large sectors of human activity because a part of this activity is deviant or abhorrent. Too many degrees of freedom are lost when general policies are written to prevent conflicts of interest.

Second, solutions to conflict of interest situations have focused on the occasion for offenses rather than on the offenders. One does not, in the United States, advocate the separation of the sexes to prevent rape, or (again in the United States) remove guns from the population in order to prevent murder. Yet, to prevent conflicts of interest, lobbyists are moved away from legislators, certain types of outsiders are prevented from organizational service, and it is suggested that relationships between regulators and regulated become more distant. A more satisfactory solution in all conflict cases might be one of cautious trust and careful screening, accompanied by the establishment of sanctions for those transgressors who, with evidence, can be found guilty of derelictions. Some conflicts of interest are improper and some are criminal. There should be laws and organizational policies directed at specific wrongdoers in conflict of interest situations. These laws and policies should identify the act as well as the guilt, and under them, wrongdoers should be punished—but with due process safeguards. Such guidelines and legislation will, however, require much more serious and detailed study of conflicts of interest than has been given to these situations in the past.

NOTES

1. Michael J. Malbin, "Political Report/New State Election Laws Add to Campaign Confusion," *National Journal,* Vol. 7 (October 18, 1975), pp. 1450–1456.
2. *Wall Street Journal,* June 12, 1975, p. 1.
3. Richard L. McAnaw, "An Adventure in Policy-Making: Ethics," paper delivered at the 1976 Annual Meeting of the Western Political Science Association, San Francisco, March 31–April 3, 1976, p. 2.
4. Plunkitt's speeches are contained in: William L. Riordan, recorder, *Plunkitt of Tammany Hall* (New York: E. P. Dutton and Co., Inc., 1963).
5. Robert K. Merton, *Social Theory and Social Structure,* revised and enlarged edition (New York: The Free Press of Glencoe, Inc., 1957), Chapter 1, especially pp. 50–54; and Philip Selznick, "Foundations of the Theory of Organization," *American Sociological Review,* Vol. XIII (February, 1948), pp. 25–35.
6. Archie B. Carroll, "A Summary of Managerial Ethics: Is Business Morality Watergate Morality?" *Business and Society Review,* No. 10 (Spring, 1975), pp. 58–60.
7. *Business Week,* March 29, 1976, p. 100.
8. Ibid.

9. It should be noted that there have been a number of court cases involving conflicts of interest by corporate board members. For example, in the Bar Chris Construction Corporation case, 1968, the court said that a lawyer on the board owed a higher duty to the shareholders than did other directors. And in *Feit* v. *Leasco,* an "outside" Leasco director was labeled an "insider" because he had helped, as a lawyer, to prepare a misleading tender offer. Theodore W. Kheel was singled out as principal in a conflict of interest case involving Stirling Homex Corporation, where he was both a director and counsel negotiating a labor contract.
10. *Forbes,* May 15, 1976, p. 118.
11. Ibid., p. 129.
12. Ibid.
13. Common Cause study reported by Associated Press, October 30, 1976.
14. Reported in *Daily Pilot,* October 30, 1976.
15. Ibid.
16. Marvin Chandler, "It's Time to Clean Up the Boardroom," *Harvard Business Review,* Vol. 53, No. 5 (September–October, 1975), p. 77, reports that only 14 percent of the corporate directors elected between January 1 and March 1, 1975, were "pseudo-outsiders" (e.g., with the potential for conflicts of interest). He feels that this percentage is still too large.
17. "Business and the House," *Economist,* Vol. 257, No. 6899 (November 15, 1975), p. 17.
18. Ibid.
19. For an interesting discussion of a specific illustration of this point, see: Eberhard Faber, "How I Lost Our Great Debate About Corporate Ethics," *Fortune,* Vol. XCIV, No. 5 (November, 1976), pp. 180–190.
20. *Life Magazine,* Vol. 51, No. 23 (December 8, 1961), p. 44.
21. Joseph W. McGuire, *Business and Society* (New York: McGraw-Hill Book Co., Inc., 1963), p. 271.
22. Raymond C. Baumhart, S. J., "How Ethical Are Businessmen?" *Harvard Business Review,* Vol. 39, No. 4 (July–August, 1961), pp. 6–8.
23. Quoted in *Newsweek,* September 1, 1975, p. 50.
24. Thomas Griffith, "Payoff Is Not Accepted Practice," *Fortune* (August, 1975), p. 125.
25. Ibid., p. 205.
26. Political Reform Act of 1974, California, Chapter 7, Section 87100.
27. "Will California's Lobbying Law Set a Trend?" *Business Week* (August 18, 1975), pp. 113–114, 116.

14

CONFLICT OF INTEREST AND PUBLIC SERVICE

EDMUND BEARD

INTRODUCTION

Public respect for government is vital to a democracy. In recent years, the American political process has suffered severe strains. A war considered immoral by much of the public proved also to be unwinnable. In prosecuting that war, elements of the American government persistently lied to its citizens. The "credibility gap" which resulted was greatly compounded by the series of scandals leading to the resignation of a Vice President of the United States under threat of criminal indictment, the conviction of members of the President's cabinet and inner circle, and finally the resignation of the President of the United States in the face of imminent impeachment. The associated investigations cast doubt upon the Department of Justice, the CIA, the FBI, the Securities and Exchange Commission, and other government agencies.

Such scandals were not limited to the Executive Branch. During much the same period, a Senator was censured for personal use of campaign funds and double billing, members of both Houses were convicted of bribery, a Senate Minority leader was accused of being on an annual $10,000 retainer from Gulf Oil, and a House Committee Chairman paraded drunkenly on a burlesque stage while another paid his mistress out of public funds.

The crisis of public confidence which resulted from these events is unmistakable. During the spring of 1974, at the height of the Watergate scandals, only 22 percent of the public expressed confidence in Congress, below even Richard Nixon's 24 percent. In 1976, the public ranked politicians 20th out of 20 occupations the Gallup Poll asked them to judge for competence and trust. Used car salesmen were 19th.

These events and observations present a clear challenge to the

American political system; we must win back the public's confidence and trust. Not only the quality of public policy but the foundation of government legitimacy is at issue. As a step in this direction, this essay will look at conflict of interest in the Executive Branch.

One observer has recently concluded that the greatest potential for conflict of interest in American politics lies not in an individual's experience or affiliation before entering government service but rather in that individual's expectations or desires regarding further private employment after leaving the government.[1] The hope of a lucrative private offer—or worse, the active solicitation of such offers—it is feared, may influence a public official's performance. Since one would hardly wish to make enemies of a desired future employer (whether through adverse judgments, careful investigative scrutiny, or stringent interpretation of the law, for example), and since the pattern of the "in-and-outer" is so widespread and accepted at both the middle and the very highest levels of American government, this is a legitimate issue. It will accordingly be treated at particular length here.

DIMENSIONS OF THE PROBLEM

In a lengthy study that played a large part in the passage of the Federal Conflict of Interest Statute of 1962, the Association of the Bar of the City of New York defined conflict of interest as the clash between

> two interests: one is the interest of the government official (and of the public) in the proper administration of his office; the other is the official's interest in his private economic affairs. A conflict of interest exists whenever these two interests clash, or appear to clash.[2]

A more recent essay by Andrew Kneier has made the useful distinction between "political interests" and "economic interests." Kneier notes that an Executive Branch official could often have a political interest in agreeing to a position advanced by industry, the White House, or some other source. Refusal could mean loss of support or access or even outright hostility. The consequent diminished political "clout" could begin to threaten one's political career. At this point, political and economic interests begin to merge. Legislators often have similar interests in voting a particular way if it means receiving or not receiving future financial or voting booth support, or if it means acceding to party loyalty or doing a little logrolling to win support elsewhere. As Kneier observes, conflicts which arise between an official's political interest and the proper performance of his duties are seldom regarded as objectionable "conflicts of interest," since virtually all federal officials face such conflicts and they are unavoidable.[3]

While political and economic interests may often overlap, however, the distinction is valuable. Setting impossible, unnecessary, or extremely unrealistic standards is hardly a solution. That conflict of interest which is to be criticized normally applies to public acts benefiting private financial interests where the conflict was avoidable and where it distorted the public acts.

Most discussions of federal conflict of interest center on three general areas: self-dealing by a public official, coupled with the more specific case of use of insider information acquired in a public capacity; assistance by public officials to private parties dealing with the government, commonly in return for some favor of economic value; and post-employment assistance by former government officials to persons dealing with the government.[4]

This listing focuses on specific, questionable actions by public officials. Yet, the New York Bar definition saw conflict of interest whenever official duties and personal financial interests clash "or appear to clash." This latter clause is quite important. In a period of growing public skepticism toward government, particularly "big" Washington government, the appearance of conflict creates as large a problem as actual conflict, since either one fosters more skepticism, more cynicism, more distrust.

WHAT ALL THIS MEANS IN PRACTICE

To the extent that a part of the current mistrust of national government is based on public perception of "conflict of interest," some might say that the appropriate solution is to change the public's mistaken perceptions rather than to promulgate new sets of "reforms" that limit officials' options and make government less attractive to able people. In this view, government officials are not nearly as bad as they are sometimes made out to be; the public mistrust of government and the suspicion that large numbers of public servants are improperly enriching themselves is inaccurate and unfair. This was the position taken by Bayless Manning when the 1962 federal conflict of interest laws were passed. Manning contended that "the record for administrative integrity achieved by modern American governments almost passes belief . . . we are living in an era of unparalleled honesty in public administration." He felt that there was no need for further regulation in the field of conflict of interest because "these restraints are no longer really designed to affect substantive behavior. They are now in large measure ritualistic."[5]

Yet, the public's fears do appear to be adequately founded. An unscientific survey of recent years reveals the following:

—An official in the Federal Maritime Commission held 589 shares of stock in a company regulated by the FMC.

—A former Assistant Secretary for Mineral Resources in the Interior Department left the Department in 1973 and became president in charge of oil shale development of Atlantic-Richfield.

—A former chairman of the Civil Aeronautics Board enjoyed a golfing vacation in Bermuda paid for by the United Aircraft Corporation at the time United was under special investigation by the CAB.

—In 1975 and 1976, the General Accounting Office found that 19 officials of the Federal Power Commission held stock in companies under FPC regulation and 159 officials in the Food and Drug Administration owned stock in companies affected by its regulations.

—In a recent four-year period, the Council on Economic Priorities found that 1,400 Defense officials left the government to join defense contractors. They included four high officials working in Minuteman procurement who were hired by Boeing, the system's prime contractor.

—Other post-employment activities include a former Environmental Protection Agency Chief who represents the polyvinyl chloride industry before the EPA; a previous Secretary of Transportation who is a lobbyist for the Union Oil Company; a former member of both the Federal Power Commission and the Interior Department who promotes oil and gas clients before the FPC. Finally, a former Secretary of the Treasury, a Secretary of HEW, and a General Manager of the Atomic Energy Commission "who helped develop and promote the plan to turn over to private industry the government's role in making atomic fuel are now executives in the Bechtel Corp., one of the firms to benefit from the new multi-billion business."[6]

—An A.P. study in the fall of 1975 found that over the previous five years, "41 top officials in regulatory agencies left to take more lucrative jobs with companies they had regulated."[7]

This brief survey touched on several different types of possible conflict of interest—from acceptance of free vacations, to stock holdings which provide the opportunity and perhaps the incentive to benefit through official decisions or "insider" information, to financially rewarding private employment upon leaving government. However, gifts to government officials can, if desired, be relatively easily defined and limited by law. Holding stock in interests "affected" by one's official duties, although presenting greater questions of definition and scope, could also be approached fairly straightforwardly. In addition, some observers feel such issues are not very troublesome for purely practical reasons. The benefits from favoring one's stock holdings, they say, are simply too risky and minimal compared to other rewards, such as deferred compensation in the form of well-paid later employment.

Post-government employment by former public officials is the area many people find most open to abuse. It is also, however, in many ways a unique problem. The dimensions of the issue are readily apparent. Government officials place contracts, some huge, in the private sector. They write, interpret, and enforce regulations. They may conduct investigations, hold hearings, air charges, propose new legislation. As noted earlier, it is hard to expect an official to do any of this energetically if it might anger or harm a desired future employer. Yet, unlike the other cases mentioned above, standards here may be much harder to define.

Should a public official have a future employment interest that may be affected by the decisions he makes (or doesn't make) on public policy? Should government agencies operate as a postgraduate residency training program in the agencies' specialties—training that then benefits the law firms or businesses that hire the young lawyer away? Should industry officials come into an agency regulating the industry and expect to return to their firms after a period of service? Should a scholar set up particular programs of contract research in an agency and then return to a university and submit a grant proposal to his former government colleagues?

On the other hand, if restraints were placed on such "in and out" movement, or if post-government service employment opportunities were limited, what would this do to the quality of public service? Could good people still be attracted to government? Is the making of public policy improved more by having access to industry expertise, if only for a time, than it is harmed by the "in and outer" going back out? What standards can be set to give us as much of the best of both worlds as possible?

Post-government employment in the private sector presents unique problems because the ties between action and reward may be hard to perceive, much less prove. The possible bias in an agency's action may not appear until years later, when an official takes a lucrative position in a firm previously favored by his rulings or judgment. The problem is particularly acute when the original favor may have been simply the absence of action or the neglect of a possible area of inquiry.

Furthermore, since the rewards for favorable official action could be both large and quite "legal"—in the form of corporate or law salaries and bonuses—the temptations will be correspondingly great. As the New York Bar Association study reported, "Interviews revealed a substantial body of opinion that government employees who anticipate leaving their agency some day are put under an inevitable pressure to impress favorably private concerns with which they officially deal."[8]

The broad dimensions of the issue have been recognized by the Committee on Legal Ethics of the District of Columbia Bar.

> If a government lawyer seeks to curry favor with a potential private employer (for example, by being overly generous in negotiations), the problem will not be solved simply by barring the lawyer from subsequently accepting private employment in the same matter. The evil is broader than that, and can be eliminated only by forbidding acceptance of private employment from any client or firm with which the attorney conducted official business while a government employee. Similarly, if a firm were to set out to hire one or more key lawyers from an agency in order to cripple the agency's enforcement efforts, disqualification of the attorney in particular matters would not be adequate to discourage such conduct.[9]

SOME SUGGESTIONS FOR CHANGE

Conflict of interest in the Executive Branch is currently regulated in several ways. Federal law is primarily designed to outlaw bribery and self-dealing, although it does place limited restrictions on post-government employment.[10] Executive Order 11222, issued by President Johnson in 1965, prohibits government officials from holding "direct or indirect financial interests that conflict substantially with their responsibilities and duties as federal employees." It goes on to state that "no employee shall solicit or accept, directly or indirectly, any gift, gratuity, favor, entertainment, loan, or any other thing of monetary value, from any person, corporation, or group" regulated by his or her agency or affected by his or her duties. The Order also states that confidential financial statements must be filed with agency heads or other designated officers.

The Civil Service Commission has specified what agencies must do to comply with federal law and Executive Order 11222. Most Executive agencies have adopted procedures prohibiting financial investments influenced by official responsibilities.

But many observers think problems remain. "The existing restrictions contain loopholes, vary from one agency to the next, and . . . are not adequately enforced."[11] Common Cause notes that "Violations of conflict of interest laws are not prosecuted by the Justice Department unless a case is referred by the Civil Service Commission or an agency head. Such cases are few and far between."[12] Even when the current "confidential" regulations are met, "the value of [a government employee's] stockholdings and other assets is not reported; nor is the amount of income received from outside activities." Directed at this system, "the investigations by the General Accounting Office . . . present a strong indictment of confidential financial reporting as a way of checking conflicts of interests."[13]

Further, the Common Cause reports note that "Presently there are no Executive Branch-wide restrictions on accepting employment after leaving an agency with a company that was affected by one's official duties." In addition, "there is no across-the-board requirement for former executive employees to report their place of employment after leaving."[14] And "a former employee may represent clients before the agency on any matter in which he was not personally involved or on any new matter which arose after he left."[15]

If problems remain, then, what can be done about them? Former Alabama Governor George Wallace, in his various bids for the Presidency, suggested dropping half of Washington's working population, weighted down by their briefcases, of course, into the Potomac River. The following pages offer some slightly less drastic proposals.

1. *Select Better People.* In recent years, several analyses have been made of presidential appointments with an eye to preventing certain problems or conflicts before they occur. The Center for Governmental Responsibility of the Holland Law Center, University of Florida, in an unpublished study found that "several problems remain. The most critical of these include a lack of integrity among some of the officials appointed in recent years, a high turnover rate, a disproportionate industry influence and an insufficiently broad search for qualified candidates." They noted the prevalence, at least prior to the Kennedy Administration, of "a method called BOG SAT, or a bunch of guys sitting around a table saying, 'Whom do you know' "?[16] Calvin MacKenzie, who has given much thought to the presidential appointment process, has found a clear improvement in the sophistication of the selection process in the Kennedy years ("a turning point in the history of the implementation of the appointment power").[17] This modernization continued, at least to some degree, in the Johnson and Nixon periods.

Whether this sophistication/modernization is an improvement may be problematic. Speaking of the Nixon years, MacKenzie states that "The President had discussed personnel requirements only in generalities. He wanted a 'broad selection of Americans' and an 'administration broadly enough based philosophically to insure a true ferment of ideas.' "[18] What actually happened was that Harry Flemming, Special Assistant for Personnel to President Nixon, and with "no previous executive branch or personnel recruitment experience . . . set up a rating system. Each applicant was assigned a code letter indicating one of the following conditions":

U—failure to appoint would result in adverse political consequences to the administration;
V—played a prominent role in the campaign, recommended in the highest terms by a member of the House or Senate leadership;
Y—no political importance to the administration;
A—applicant not compatible with the Nixon Administration and should not be appointed to public office.[19]

Of course, political agreement, political effectiveness, and political expedience should and will influence certain presidential appointments.[20] To make personal or political loyalty the only standard, however, may prove disastrous. The Center for Governmental Responsibility concluded that the President needs an organized, thorough search for superior talent if the quality of the government's executive officers is to be substantially improved. It recommended specifically that an office of presidential appointments be created to conduct a broad search for highly qualified individuals; release the names of major candidates for public comment; see that logs are kept of contacts from outside parties concerning the selection of appointees; and require public financial disclosure by and security and financial checks on potential nominees. In addition, nominees should commit themselves to serve for at least two years.[21] After leaving public service, appointees should be prohibited from representing an outside party before the agency in which they served for two years, this subject to specific civil or criminal penalties.[22]

2. *Financial Disclosure.* Working from the perspective of the Senate and its confirmation power, Judith Parris has specified what Senate Committees might require in the way of financial disclosure of nominees.

In order to assure the absence of conflicts of interest, all individual nominees should be required to elaborate their financial situation. They should submit a list of all the property, real and personal, and all the debts and other financial obligations of themselves and the members of their immediate families. They should also set forth the sources and amounts of all income received during the past several years, including income listed by clients for professional services. Any anticipated or deferred income also should be listed and the source indicated. Past business relationships with the federal government, any lobbying activity, and any other financial dealings that might be considered conflict of interest should be elaborated, along with the nominee's proposed means of resolving any conflict. Any plans regarding future employment relationships also should be set forth in detail.[23]

William Proxmire, Chairman of the Senate Banking and Currency Committee, tried at the beginning of the 94th Congress to get the Democratic caucus to adopt an extensive questionnaire for all nominees along the Parris model. Although the idea was not accepted for all committees, several, including Banking, did agree to use it. Of course, the specificity and the accessibility of any such report are crucial to its effectiveness.

3. *Divestiture*. The most extensive set of recent recommendations regarding conflict of interest in the federal government has come in a series of reports from Common Cause. Their recommendations have been aimed at four results: public financial disclosure by high-level executive personnel to remedy the deficiencies in confidential reporting; divestiture of financial interests that conflict with public duties which would significantly tighten the current loosely drawn requirements; restrictions on post-government employment with interests that were dealt with while in office; and improved review and enforcement mechanisms.

The Common Cause recommendations set "a strong presumption in favor of divestiture of any financial interests that conflict with one's duties. Any agency employee who participates "personally and substantially" in an agency proceeding or a policy proceeding must divest of all financial interests in any "person" affected. In the case of a policy proceeding, a substantial part of the "person's" activities must be regulated or otherwise affected by the agency in question. There is a procedure for the granting of exemptions to these requirements if agency counselors or special review committees recognize "special circumstances . . . [which] would impose unfair burdens or jeopardize the achievement of overriding public interests." The facts behind, and an explanation of, such a judgment must be made public.[24]

4. *Restrictions on Post-Government Employment*. The most interesting proposal put forth by Common Cause deals with post-government employment restrictions. They suggest that federal employees enter into legally binding contracts that prohibit, for two years after government service:

—accepting employment with a company or organization that was affected as a party to a specific agency proceeding in which they participated;

—accepting employment with a company or organization that was affected by a policy proceeding in which they participated, provided the entity was substantially regulated by their former agency; and

—representing any party before their former agency in any legal, lobbying, or other professional capacity.[25]

Ex-government employees would also be required to report for two years their current occupation and place of employment. These would be made public. Any arrangements for future employment while in government service would also be reported, and engaging in any agency action affecting the future employer would be prohibited.[26]

These rules would extend, both in time and coverage, the requirements of current law. Enforcement would be accomplished through review of the financial statements by the Civil Service Commission and designated agency officers. Written notification of the results of the review would be required. The public would have access to the reports. Civil proceedings could be initiated to force compliance with contracts regarding post-employment. Sanctions while in office would include termination or temporary suspension from the agency. The courts could order compliance in cases of post-employment controversy.

5. *Enforce the Present Laws.* The lack of specificity in current regulations, the "confidentiality" of what reports must be filed, and either the non-existence of enforcement offices or their failure to act, have all been referred to above. Specific examples may make the situation clearer. The Council on Economic Priorities recently reviewed 1,400 high-ranking persons who had left the Defense Department and accepted private employment with defense contractors since 1969. Current law requires that former Department of Defense personnel file a yearly report on where they are employed. Another law sets a lifetime prohibition on retired military officers selling materials to their former branch of the armed services. Yet, the Council on Economic Priorities found that "no agency is assigned the enforcement of the reporting requirement." The CEP estimates that one-third to one-half of ex-military officers who should have filed did not. "There has never been a prosecution for failure to report." Furthermore, the no-selling-to-your-ex-service law "is vague, contains no express criminal penalties, and has not been enforced at any time by those agencies responsible for criminal-law enforcement. Nor has there been a judicial test of the adequacy of the law."[27]

6. *Raise Salaries.* One other proposal to alleviate many of the problems of conflict of interest is to raise government salaries. The reasoning is simply that if the government servant is earning "enough" money, he or she will be less tempted to act improperly to acquire even more. A decade ago, Dean Mann noted that "inadequate compensation is still a deterrent to acceptance to political positions. It is also a factor in the decisions of many men to leave the government."[28]

The Center for Governmental Responsibility has added that "Despite the pay raises from 1964 through 1969, the problem still remains. In the last seven years, the cost of living has risen 50%, while the pay for the highest 786 government officials has risen only 5%. At the same time, executive salaries in the business world have increased on the average 60%, more than compensating for inflation."[29] Stephen Hess has recently suggested that

> Members of Congress should be paid a salary sufficient to insure their exclusive services. When national legislators are compensated at the level of a Cabinet Member—$60,000 a year—we should then have every right to require, as a condition of employment, that they sever all outside business and law interests, that their investments be placed in a blind trust, and that their time and energy be devoted completely to the public's business. The tradeoff is that we assure their financial independence in return for the guarantee that their actions will not be influenced by possibilities of private gain that may result from their exercising public duties.[30]

Hess would presumably apply the same argument to Executive Branch officials. The problem that many of these suggestions ignore (or settle to their own but perhaps not their subjects' satisfaction) is simply, "How much is enough?" Why is $60,000 adequate to restrain a person's greed? It could as well be argued that a person who can't live on $30,000 or $40,000 a year will be precisely the person who thinks he can't live on $60,000. Certainly the question is not easily settled. In recent memory we have had cabinet officers, a Supreme Court Justice, a Vice President, and a President who felt the urge to supplement their large official salaries.

7. *Select Better People (revisited).* It is possible that the narrow spectrum of Americans from which Executive Branch appointments are made tends to accentuate certain conflict of interest problems. If salaries and other rewards are much higher in the business world, persons accustomed to that world may chafe in government service. Raising salaries might limit this, but possibly only if the salaries were raised

quite high—undoubtedly high enough to create another problem in taxpayer discontent.

The advantages of widening the search process for appointees, of course, lie not simply in the hopes of finding able persons with lower salary expectations. Indeed, there is no reason to assume that professors, or state and local public officials, or labor union leaders, or others are more honest or less susceptible to temptation than businessmen. The reverse might be argued—that professors could see public service as their one chance to get rich, or that the temptations would appear even greater to persons used to lower incomes. A much more important reason for widening the search is to broaden the range of perspectives brought into government.

The Center for Governmental Responsibility has recently observed that "the result of these procedures [of recent presidential appointments] has been to produce a preponderance of industry representation and a dearth of representation from other segments of the interested public. . . . The appointment procedures do not appear to have reached very far into all sectors of American life to seek out potential talent for public office."[31]

As just one example of this phenomenon, a Common Cause study of the employment backgrounds of high officials in the Energy Research and Development Administration (ERDA) and the Nuclear Regulatory Commission (NRC)—the two heirs to the Atomic Energy Commission—found that

—52% of top ERDA employees came from private enterprises involved in energy activities; 75% of these came from ERDA contractors;

—58% of top NRC officials came from private enterprises holding licenses, permits, or contracts from NRC;

—88% of the NRC consultants who are university professors or employees of other non-profit institutions (e.g., research labs) come from institutions regulated by the NRC;

—Six of the nine Executive-Level positions surveyed in ERDA, including that of Administrator, are filled by individuals previously employed by commercial firms who are today ERDA contractors.[32]

A pro-industry bias simply in the ways problems are perceived, structured, and approached may thus be inevitable. In former FCC Commissioner Nicholas Johnson's words,

> If a person goes into industry, acquires friends in industry, goes to industry parties and conventions, sees his whole life-style determined by the vicissitudes of that industry, [and] acquires knowledge about the industry along the way, the paramount learning experience is not the subject matter but rather, the social framework in which it is acquired.[33]

This pattern is often repeated when "advisory councils" or groups are set up to work with the agency. As a Ralph Nader study found in 1970, "The ICC has numerous advisory groups to help set policy at the initial, formative stages within the agency—including not one consumer or consumer representative."[34]

Although the composition of such outside advisory councils can and should be more balanced, it is important to keep in mind the realities of the situation. In the words of a recent excellent study of presidential appointments to the Federal Communications Commission and the Federal Trade Commission, "Avoidance of the controversial nominee, coupled with stout opposition in influential quarters from the regulated interests, almost assures that no individual with a background in the consumer movement will be seriously considered for appointment."[35]

CONCLUSIONS

This essay is not an inclusive survey of all real or imagined ills in government. Nor is it a diatribe against specific abuses or specific persons accused of committing them. It has tried to define certain dimensions of the conflict of interest problem and has described some of the proposed reforms. It has also identified the dilemmas inherent in this issue and the tradeoffs involved. For the most part, value judgments have been avoided and sides have not been taken. Departures from this impartiality are simply the result of human frailty—which, of course, could also be said of the questions here discussed.

Some of the problems noted here would be alleviated by a wider personnel search, disclosure, required divestiture, or higher salaries. Others might be controlled by more effective enforcement of laws already enacted. All of them could be eliminated by doing away with government. They could also be controlled by public opinion and public voting—if that public was in agreement and focused.

Most observers think the biggest problem in conflict of interest lies in employment after government service. The "in and outer" has been often mentioned here. Yet, certain aspects of this issue have been obscured. The "in and outer" is, in many respects, the essence of democratic government. The benefits from longer tenure of public officials seem obvious: more consistency, more stability, more experience, less waste from constant retraining and reworking. The "independence" and "dependability" of the British civil service is often presented as a model. The American "in and out" pattern certainly presents special problems. It is inefficient and sometimes suspect. But so is democracy. This is, after all, meant to be a "government of the people." The trick is to get the least expensive, most effective, most impartial, most wise, and least corruptible government. If the answer is not immediately obvious, the quest is still vital.

NOTES

1. C. MacKenzie, "The Appointment Process: The Selection and Confirmation of Federal Political Executives," Ph.D. dissertation (Cambridge, Mass.: Harvard University, 1975), pp. 110–111.
2. Association of the Bar of the City of New York, *Conflict of Interest and Federal Service* (Cambridge, Mass.: Harvard University Press, 1960), p. 3.
3. Andrew Kneier, "Ethics in Government Service," in *The Ethical Basis of Economic Freedom,* ed. by Ivan Hill (Chapel Hill, N.C.: American Viewpoint, Inc., 1976), pp. 217–218.
4. See, for example, Roswell B. Perkins, "The New Federal Conflict of Interest Laws," *Harvard Law Review,* Vol. 76 (April, 1963), pp. 1118–1119.
5. Bayless Manning, "The Purity Potlatch: An Essay on Conflicts of Interest, American Government, and Moral Escalation," *Federal Bar Journal,* Vol. 24, No. 3 (Summer, 1964), pp. 241–248.
6. These examples have been drawn from Andrew Kneier, "Ethics in Government Service." The quoted line is found on page 226.
7. "Presidential Appointments," Center for Governmental Responsibility, University of Florida, July 1976, p. 10. (Hereafter cited as "Presidential Appointments.")
8. *Conflict of Interest and Federal Service,* p. 233.
9. Committee on Legal Ethics, District of Columbia Bar, July 12, 1976, "Inquiry No. 19, Tentative Draft for Comment," p. 15. (Hereafter cited as "Inquiry No. 19.")
10. 18 *USC,* 201–208.
11. Kneier, "Ethics in Government Service," p. 222.
12. "Draft Introduction" to "Conflict of Interest Report," Common Cause, Summer 1976, p. 8.
13. "Draft Introduction," p. 3.
14. Although this is required by the Defense Department.

15. "Draft Introduction," pp. 5–6.
16. "Presidential Appointments," p. 1.
17. MacKenzie, "The Appointment Process," p. 93.
18. Ibid., p. 106.
19. "Presidential Appointments," p. 6.
20. A useful recent article defending this position is Charles Peters, "A Kind Word for the Spoils System," *Washington Monthly* (September, 1976), pp. 26–31.
21. In looking at eighty-four Eisenhower and Truman appointees, Dean Mann discovered that the average service in office was slightly over two years. "Only thirteen had a position tenure of four years or more and there were eight who left before twelve months were up." [Dean Mann, *The Assistant Secretaries* (Washington, D.C.: The Brookings Institution, 1965), p. 228.] A panel of the National Academy of Public Administration noted the same tendency in recent years in testimony before the special Senate Watergate Committee. See "Watergate: Its Implications for Responsible Government," A Report Prepared by the National Academy of Public Administration, *Administration and Society,* Vol. 6, No. 2 (August, 1974), pp. 155–170.
22. "Presidential Appointments," Summary of Recommendations, pp. 25–26.
23. Judith Parris, draft manuscript on the Senate confirmation process.
24. See Common Cause, "Model Executive Order on Conflict of Interest," a separate "Summary" of this Model Order, and the Draft Introduction to the "Conflict of Interest Report," cited above, all Summer, 1976.
25. Common Cause, "Summary" of "Model Executive Order on Conflict of Interest," Summer, 1976, pp. 3–4.
26. Ibid.
27. Leon S. Reed, "Military Maneuvers" (Washington, D.C.: Council on Economic Priorities, 1975), pp. 3–4. By their not unreasonable standards, the CEP found "379 (or 27 percent) of the 1,406 former DOD officials and retired officers . . . [who] met one or more of these criteria for conflicts of interest" (p. 2).
28. Mann, *The Assistant Secretaries,* pp. 278–279.
29. "Presidential Appointments," p. 14. The CGR cites *Congressional Quarterly,* June 12, 1976, p. 1515.
30. *The Wall Street Journal,* July 9, 1976.
31. "Presidential Appointments," p. 13.
32. Common Cause, "Employment Backgrounds of ERDA and NRC Officials," June, 1976, p. 2.
33. James Graham, and Victor Kramer, *Appointments to the Regulatory Agencies,* printed for the Committee on Commerce, U.S. Senate (Washington, D.C.: Government Printing Office, 1976), p. 398.
34. Robert Fellmeth, *The Interstate Commerce Omission* (New York: Grossman Publishers, 1970), pp. 312–313. This study also found that

"Job interchange levels between the ICC and industry have grown, with 'deferred bribes' becoming the norm. Many officials admit they receive job offers from industry while in government employ. In the past decade, all but one Commissioner who has left the agency has ended up working for a carrier or a carrier association directly, or indirectly as an ICC Practitioner."

35. *Appointments to the Regulatory Agencies,* p. 400.

15

OPPORTUNITIES FOR SELF-ENFORCEMENT OF CODES OF CONDUCT:

A Consideration of Legal Limitations

EARL W. KINTNER
and ROBERT W. GREEN

I. THE NEED FOR SELF-ENFORCEMENT

Before turning to a consideration of how self-enforcement of codes of conduct may be effected and what legal limitations may exist in this area, it is appropriate to disclose briefly certain fundamental assumptions that underly this essay's assertion of a strong public interest in the development and proper self-enforcement of such codes.

A pragmatic inquiry into law enforcement efforts undertaken by the government clearly suggests that such efforts are never so pervasive as to reach all instances of unlawful behavior. Moreover, attempts to increase law enforcement activities in an effort to discover and punish such unlawful behavior must inevitably bring with it an increasing governmental presence and a greater governmental intrusion into the fabric of our society. In short, allowing all enforcement responsibilities to devolve upon the government could ultimately produce authoritarianism—a society in which every significant activity would be subject to direct government regulation. Such a development would be antithetical to the democratic precepts upon which this nation was founded.

Fortunately, there is a clear alternative to authoritarianism: recourse to the strength and initiative of the citizenry itself. Laws and law enforcement develop from those values and mores which the society has internalized. Adherence to these societal values was originally produced through the application of peer pressures and the moral suasion of the group as a whole, perhaps as interpreted by its leaders.

Creation of written laws and formal enforcement methodologies represent a later development in establishing social controls.

While society may be regulated either by the moral suasion of the group or through a formal enforcement activity, the former approach has the obvious virtue of being more effective. It remains in operation at all times and at all levels within the society and is capable of addressing virtually all significant deviations from accepted social norms. The second approach, whose use is based upon an assumption that a formal sanctioning entity separate and apart from the rest of the society is best, is necessarily limited by the financial and manpower resources made available to it, and must establish priorities as to how those resources are to be applied.

Problems quickly arise when members of the society begin to rely solely on the formal law enforcement entities to enforce and protect values of the society at large. Given the fairly limited extent of the economic resources which society is willing to devote to the law enforcement function, it is clear that the formal enforcement agency cannot possibly detect and prosecute every significant deviation from the law. One possible response to these practical limitations—a response which has found considerable Congressional favor during the past ten years—is the creation of broad new laws directed specifically at certain types of behavior and the creation of new agencies to enforce these laws.

This approach to law enforcement sets up a vicious circle: the broader laws and newer regulatory agencies will inevitably uncover an increasing number of law violations, and this discovery will, in turn, engender still broader laws and regulations and even wider enforcement activities. If this trend is carried to its logical conclusion, it will produce the sort of iron statutes which will eliminate the flexibility that is necessary if the law is to be effective and its enforcement just and nondiscriminatory. Neither personal liberty nor commercial intercourse can prosper under such laws.

In addition to offering a clearly superior method for providing sanctions for unlawful behavior, codes of conduct can do much more to support and foster general social values through their ability to stimulate the education and self-regulation of industry members. Even without provisions for self-enforcement, codes of conduct serve to identify and encourage desirable activities by formally establishing a high standard against which each individual or company may measure its performance. Likewise, existence and knowledge of these high standards serves to discourage undesirable and unlawful activities by identifying conduct which is unacceptable to the group. Appropriate and effective codes of conduct can eliminate or greatly minimize the need for the sorts of pervasive government regulation which may unduly limit individual freedoms and destroy individual initiative and the private enterprise system.

Viewed from this perspective, effective codes of conduct are not merely useful and desirable; they are absolutely essential if our democratic society is to reach its greatest potential. Moreover, if such codes are to achieve their maximum contribution, there must be a method by which adherence to such codes may be enforced. Starting from the premise that government enforcement of such codes would produce precisely the sorts of additional governmental intrusion into the social fabric that we are trying to avoid, the need for self-enforcement of these codes is apparent. Fortunately, the extensive use of codes of conduct for beneficial social purposes is already permitted under present law, and there is a growing recognition that the private enforcement of such codes is appropriate in our democratic society.

At the same time, of course, we recognize that codes of conduct and their enforcement do not exist in a vacuum. Consideration of the public interest must always be paramount in the establishment, publication, and enforcement of the codes of conduct. Most of the litigated cases involving the appropriateness and legality of codes of conduct involve instances in which inappropriate provisions were included in the code or situations in which the best interests of the industry members were given preference to the public interest. The following discussion of these cases is intended to set up the permissible scope for codes of conduct and delineates the extent to which the law permits self-enforcement of such codes.

II. PERMISSIBLE SUBJECTS FOR INCLUSION IN CODES OF CONDUCT

Although historically a code of conduct developed as a covenant among peers, modern codes of conduct must be based as well upon integral considerations of the public interest—especially with regard to those laws regulating trade and commercial relationships. In short, it is not enough that a code of conduct have substantial beneficial effects both for the group which developed it and for the general public. It must not achieve those beneficial effects through provisions which are in and of themselves unlawful. For example, provisions in a code which unduly limit competition or unreasonably restrain trade or commerce in a particular business or industry affirmatively transgress the prohibitions of the antitrust laws and expose those who develop and observe the code to potential civil liability and criminal prosecution.

Moreover, the governmental scheme established by the United States Constitution does not contemplate or permit the delegation of legislative power by the Congress to private groups. Accordingly, there are limitations upon the sorts of behavior for which self-enforcement may be permissible. As the later discussion will point out, it is one thing for a code of conduct to provide enforcement sanctions directed at acts which constitute violations of existing laws. It is quite another thing, however, for a private group to arrogate legislative authority to itself by declaring additional acts to be unlawful and sanctioning those who commit such acts.

Finally, a clear distinction must be made between codes of conduct or codes of ethics, on the one hand, and voluntary industry product or service standards, on the other. While voluntary standardization efforts can play a vital role in the improvement of product and service quality, such activities involve—at least tacitly—the agreement of direct competitors concerning the character, quality, or composition of the product or service in question, and such activities are measured against far more stringent antitrust standards. Accordingly, such industry standardization efforts should be developed separate and apart from any code of conduct which establishes the general ethical precepts to which a particular group subscribes. Because of the complex factors which determine the legality of privately developed product and service standards, no further discussion of these areas has been attempted in this essay.

A. Compliance with the Antitrust Laws

Acting to correct the monopolistic practices of the great commercial trusts, Congress in 1890 passed the Sherman Act, which declares in simple language that "[e]very contract, combination . . . or conspiracy, in restraint of trade or commerce among the several States, or with foreign nations, is hereby declared to be illegal." This language produced some initial enforcement difficulties, since every contract between business entities necessarily entails certain restraints upon the freedom of both parties to act with respect to that contract. Twenty-one years later, in *Standard Oil Co.* v. *United States*, 221 U.S. 1 (1911), the Supreme Court clarified this language by noting that these prohibitions were to be applied under a "rule of reason" and that only those contracts, combinations, or conspiracies which are found on their particular facts to be unreasonable are proscribed by the Sherman Act. Although there have been a number of subsequent amendments and additions to the antitrust laws, section 1 of the Sherman Act (quoted above) remains unchanged and provides the touchstone against which legality of the provisions of codes of conduct must be measured.

With the exception of certain types of commercial restraints which the courts have found to be so pernicious and destructive of competition as to foreclose any need to consider whether they might be reasonable, the rule of reason inquiry continues to be the applicable test for violations of the Sherman Act. Included within the limited number of practices which have been found so devoid of valid competitive purpose as to preclude any need for consideration under the rule of reason are agreements to fix prices, agreements among competitors to allocate geographic territories or customers, and the use of "tying" contracts in which a buyer who wishes to purchase one commodity is also required to purchase a second commodity which he did not want. These practices constitute per se violations of the antitrust laws, foreclosing any judicial inquiry as to either their purpose or effect. Once the existence of the particular practice is proven, judgment against the party using the practice follows as a matter of law.

Although the Sherman Act has proven to be otherwise highly flexible in its application and capable of reaching and controlling types of activities never contemplated by its draftsmen in 1890, on the per se violation side this law operates as an iron statute. It is legally immaterial whether a particular trade practice might, if analyzed in depth on its particular fact situation, be found to be reasonable within the purview of the antitrust laws. If the behavior in question falls into one of the judicially proscribed per se categories, the court is *not permitted* to engage in any independent inquiry as to the legality of the practice. Neither favorable competitive facts nor highly laudable intentions of the defendants are material to the court's decision.

Assuming that a proposed code of conduct scrupulously avoids any provisions which are per se violative of the antitrust laws, the effect of the code's provisions on the conduct of any trade or business will continue to be tested under the rule of reason. Since both the direct and indirect effects of a particular provision proposed for inclusion in the code are subject to antitrust inquiries, it is imperative that each provision be separately examined to be sure that it is reasonable under the circumstances.

B. Ethical Precepts Versus Competitive Restraints

As the foregoing discussion suggests, each provision being considered for inclusion in a code of conduct must be subjected to a critical and objective analysis. Great care must be taken to distinguish between those proposed provisions which are based upon ethical precepts and those which relate primarily to establishing particular trade customs. Provisions based upon ethical precepts will generally pass muster under the antitrust laws, but provisions involving restrictions upon competitive activity will almost invariably raise problems.

In general, ethical precepts are founded upon a concern for doing the right and proper thing in one's behavior toward others. In this context, such general absolute concepts as honesty and fair dealing almost always will be embraced in the development of a code of conduct. Incorporation of such precepts is totally appropriate and necessary to give the code practical vitality.

While these general ethical precepts are desirable and clearly unobjectionable, any further attempts by a group of competitors to identify and specify particular commercial practices which they would collectively like to see eliminated can raise a substantial antitrust problem. For example, a group of merchants with stores located in a shopping center might propose prohibitions against making retail sales on Sundays or might propose to prohibit the retail sales of any commodity or service below cost, in an ostensible effort to eliminate unfair competition with other merchants dealing in the same commodities or services. In both cases, however, it is clear that the retailers are not primarily concerned with the fair dealing with the consumer but are seeking impermissibly to standardize hours of operation or place an absolute lower limit on the prices at which goods will be sold to the consumer. Provisions such as these are naked restraints of trade that will not successfully withstand antitrust challenge.

Since codes of conduct are typically developed by individuals who are in competition with each other, whether in the sale of goods or services or in the provision of professional services, the distinction between ethical precepts and competitive restraints must be scrupulously observed. While there are a number of forms of competitive behavior which have been determined either to be per se unlawful or which have been determined to be unlawful in particular cases by the courts, there are many other types of vigorous competitive activities which are neither unfair, deceptive, nor unlawful, and any combined effort to eliminate these practices will be patently unlawful. In general, when certain commercial practices are not dishonest or deceptive, do not take unfair advantage of the consumer, and have not been determined by the courts or the Federal Trade Commission to be unlawful, it is likely that attempts by competitors to eliminate such practices will not be lawful. In any case in which doubt exists, careful legal review of proposed code provisions is a necessary precaution.

In order to illustrate some of the kinds of problems which particular restraints on competition have been held to raise, the following brief discussion has been provided. These cases establish, among other things, that the good intention of the group in selecting particular provisions for its code is no defense to an antitrust prosecution.

C. Unlawful Code Provisions*

Code of conduct provisions which directly or indirectly fix or stabilize prices are the most vulnerable area under the antitrust laws. Regardless of the good intent of a particular provision, including the elimination of certain unequivocally unfair or deceptive practices, if the provision also has a direct effect upon the prices of goods or services it will not pass antitrust muster. *Standard Sanitary Manufacturing Co.* v. *United States,* 226 U.S. 20 (1912), is an early case establishing this principle. The Association of Sanitary Enameled Ware Manufacturers purchased the dominant patent under which such products were manufactured and then licensed all industry members on the condition that they adhere to certain price schedules and not market "seconds."

Although such seconds were allegedly being palmed off on the public as first-quality goods and customers were being deceived thereby, the primary purpose of these activities was to enhance industry conditions and to eliminate price competition by withdrawing substandard products from the market. On review, the Supreme Court found that elimination of price competition among the companies and elimination of price competition through the marketing of lower-priced seconds was clearly anticompetitive, could not be justified, and was illegal, notwithstanding any possible beneficial effect to the industry flowing from the code.

* Most of the following case law discussion has been adapted from Kintner, "Legal Limitations and Possibilities for Self-Enforcement of Codes of Ethics," in *The Ethical Basis of Economic Freedom* (I. Hill, ed., 1976), copyright 1976 American Viewpoint, Inc., Chapel Hill, N.C., and is reproduced herein by permission of the publisher.

In a later case, existence of an association code of ethics providing that, among other things "sugar should be sold only upon prices and terms publicly announced," was accepted as evidence of an illegal price agreement by the Supreme Court in *Sugar Institute* v. *United States*, 297 U.S. 553 (1936). Likewise, existence of an association purpose to maintain a multiple basing-point delivered-price system was accepted as part of the evidence establishing an unlawful price agreement in *FTC* v. *Cement Institute*, 333 U.S. 683 (1948).

To the same effect, the Federal Trade Commission has issued a number of advisory opinions dealing with association codes and the legality of including certain provisions with indirect price-restricting consequences. For example, a wholesaler group's proposed resolution establishing uniform prices and terms to be paid to its members for worn-out parts by their parts rebuilder customers was refused approval in Advisory Opinion Digest No. 15. The following provisions have also been explicitly disapproved: establishment of uniform terms to govern dealing with commercial customers (Advisory Opinion Digest No. 97), provisions calling for "fair and adequate profit levels" (Advisory Opinion Digest No. 115), provisions seeking firm price quotations from suppliers (Advisory Opinion Digest No. 137), provisions prohibiting sales below cost (Advisory Opinion Digest No. 249), and prohibitions against the advertising of rates for performing services (Advisory Opinion Digest No. 268).

Like price-fixing provisions, group boycotts receive invariably harsh treatment under the antitrust laws. For example, the attempts of retail lumber dealers to eliminate direct retail sales by lumber wholesalers through the circulation among the trade of lists naming such direct selling wholesalers was declared to be a violation of the Sherman Act in *Eastern States Retail Lumber Dealers' Association* v. *United States*, 234 U.S. 600 (1914).

Even when a boycott is aimed at eliminating questionable competitive practices, it is unlikely to be held lawful under the antitrust laws. For example, in *Fashion Originators' Guild of America, Inc.* v. *FTC*, 312 U.S. 457 (1941), the Supreme Court upheld the Commission's determination that an organized boycott established by fabric manufacturers in an attempt to eliminate so-called style pirates from the market was unlawful. The Court's opinion clearly implies that the only legitimate source of protection for fabric designs must lie in the copyright and patent laws, and it describes the Guild as being "in reality an extra-governmental agency, which prescribes rules for the regulation and restraint of interstate commerce, and provides extra-judicial tribunals for determination and punishment of violations and thus 'trenches upon the power of the national legislature and violates the statute'" (312 U.S. at 465).

Two years later, in *U.S.* v. *American Medical Association*, 317 U.S. 519 (1943), the Supreme Court, in ruling upon the AMA's efforts to boycott doctors who practiced for or consulted with a plan offering prepaid medical care, rejected the argument that such professional groups, which under state law have been given some authority to control aspects of professional practice, are thereby exempt from the strictures of the antitrust laws. In 1975 the Supreme Court reinforced this opinion by holding that there is no professional exemption available to a bar association seeking to justify its establishment of uniform minimum fee schedules (*Goldfarb* v. *Virginia State Bar*, 421 U.S. 773 [1975]).

Virtually all of the unlawful practices outlined above are activities which do not properly belong in a code of conduct anyway. They are detailed here solely as a catalogue of provisions which should be rigorously avoided in formulating a good and effective code of conduct. Inclusion of any one of these practices can later result in an otherwise exemplary code of conduct being used as evidence that a group conspired to violate the antitrust laws.

D. Lawful Code Provisions

Having catalogued some of the ways in which codes of conduct can violate the law, it is only fair to recognize that such codes can also make a preeminent contribution to the well-being of society and can encourage a finer and more desirable quality of life for everyone. Nothing in a proper code of conduct or in its appropriate enforcement is inherently violative of either the antitrust laws, the general laws, or the mores of our society. In fact, it is questionable whether our democracy can continue to exist unless the behavior of its citizens is guided by proper ethical precepts.

For example, there are many types of business practices which have been ruled unlawful by the courts or the Federal Trade Commission, but neither the Commission nor the Department of Justice has the resources, manpower, or inclination to seek out each of these practices wherever it may be occurring in our economy. Here is the area where industry cooperation and a code of ethics can go a long way toward eliminating the more undesirable of these practices in its own industry. A good place to begin is in the area of advertising, establishing standards for honesty and full disclosure of material information concerning the industry's products. Far from imposing any restraint on trade, the elimination of false and misleading advertising has a strongly procompetitive effect; it ensures that informed consumers will be enabled to make rational choices in their selection of goods or services.

Self-regulation of advertising practices is, in fact, an important function of many viable industry codes. Moreover, most of the situations in which either the FTC or the Justice Department has approved self-enforcement of an industry code have involved code provisions designed to eliminate deceptive practices in advertising or other representations made to consumers. For example, prior to the establishment of federal legislation limiting the broadcast advertising and other advertising of cigarettes, the industry had developed and received approval from the Justice Department of a comprehensive Cigarette Advertising Code, in which certain standards were laid down governing the content of cigarette advertising. Under the Code, all cigarette advertising was required to be submitted to an independent administrator for advance clearance, and the Code contained provisions requiring the payment of a sum of up to $100,000 as liquidated damages for each violation. A key element of this Code was the strict limitation upon the promotion of cigarettes on school or college campuses and the prohibition against using free samples to solicit trade of persons under twenty-one years of age.

As a practical matter, the individual cigarette companies would have found it almost impossible individually to adhere to the Code's standards without the assurance that their competitors would also follow the same standards. Similar codes dealing with program and film content have been established in the motion picture, radio, and television industries, permitting these industries as a whole to achieve higher standards than any one of them could have achieved alone.

In this age of consumerism, no industry can with impunity ignore general industry practices which have a capacity to deceive or mislead consumers. With the increased budgets of the Justice Department and the FTC, and a growing militance among consumers who feel that they have been dealt with unfairly, chances are excellent that one or more companies in the industry will be selected and prosecuted to establish the illegality of these practices. Typically, the government selects only one company and makes an example of it, since the cost of proceeding against all members in the industry would be prohibitive. Considering the severe competitive handicap which an adjudicated case can impose upon the company selected for enforcement activity, development of and adherence to appropriate industry advertising standards is a far more intelligent course, permitting each company's attention to be directed to enhancing its competitive position rather than defending itself in a lawsuit.

Where certain practices within an industry are pernicious enough and self-enforcement activities alone have not been sufficient to eliminate harmful practices, Congress has occasionally acted by creating a new regulatory commission with plenary powers to impose uniform standards and requirements upon an industry. The enactment of federal legislation dealing with automobile safety in 1966 and legislation dealing with consumer product safety in 1972 can be directly traced to perceived shortcomings in the efforts of industries to regulate themselves. Needless to say, these agencies and others like them have provided substantial additional burdens upon the industries affected and have greatly complicated the conduct of their businesses.

In 1967, a group of magazine publishers who were concerned about certain abuses in the door-to-door sales of magazine subscriptions secured FTC approval for a self-regulatory program designed to identify salesmen guilty of deceptive practices and publicize their names to all subscription agencies employing such salesmen (Advisory Opinion Digest No. 128). In addition to identification of offending salesmen, the approved Code also provided for penalties of up to $5,000 per violation to be imposed upon subscription agencies violating the provisions of the Code. All enforcement activities were to be handled by an independent administrator retained to enforce the Code's provisions. The Commission expressed its judgment that such fines "will not operate anticompetitively or in a confiscatory manner but [will be] sufficient to constitute a deterrent" (letter from Federal Trade Commission Secretary Joseph W. Shea to Earl W. Kintner, dated May 22, 1967). Two further conditions imposed by the Commission deserve note: first, the requirement that participation in the Code by subscription agencies was to be on a completely voluntary basis, and second, that approval was granted for a three-year period subject to reconsideration by the Commission at the end of that time. The plan expired in 1970 and no request was made to renew it, since the FTC had in the meantime issued formal complaints against a number of participants concerning alleged misrepresentations by salesmen in the sale of subscriptions.

No subsequent FTC advisory opinions concerning specific self-enforcement programs have been issued, but in 1969 and again in 1974, the Commission issued advisory opinions recognizing the appropriateness of expulsion from association membership for those members refusing to adhere to code provisions modeled upon requirements imposed by law. In the first case, the FTC gave its explicit approval to the enforcement by an association of shippers' agents of Code provisions requiring member compliance with legal limitations imposed by the Interstate Commerce Act on the scope of permissible shipping activity (Advisory Opinion Digest No. 373). Under this Code, repeated failure to comply would, subject to notice and an opportunity for a hearing, result in either a probationary period or expulsion of the offending member. The Commission's opinion notes:

> While the Code contains provisions restricting the business operations of members, it appears from the materials submitted that the purpose of these restrictions is to insure that members remain within the Act's limitations and respect the confidential agency status created in their dealings with shipper-customers. The purpose is also to encourage Association members voluntarily to refrain from unfair or deceptive practices. In this context there is a greater public interest in protecting shippers from dishonest shippers' agents than there is in condemning the minimal restraints that might result from application of the Code.
>
> Undoubtedly, unreasonable and therefore unlawful restraints might result if an Association member is arbitrarily or improperly expelled from membership, but the Commission believes that there is ample public interest in effectively encouraging Association members to refrain from the clearly pernicious practices condemned by the Code. On the assumption that the Code will be administered in such a way as to promote this end, and not so as to place unreasonable restraint on the ability of members to do business, the provision permitting the Association to expel non-conforming members is approved. (76 FTC 1108)

Similarly, in 1974 the Commission issued an unnumbered advisory opinion to the Constitution and Bylaws Modeling Association of America, International, concerning a revised draft of that Association's constitution and bylaws (*Trade Reg. Rep.* ¶20,549, FTC, March 20, 1974). While declining to approve the specific language proposed by the requestor concerning the basis for dismissal of a member, the Commission stated the following criterion for provisions dealing with expulsion: "Dismissal from membership should be allowed only for failure to comply with specific, non-discriminatory, objective criteria that adhere closely to the requirements of the law." The advisory opinion further suggested that any new language selected by the respondent could again be submitted to the Commission for clearance as to its legality under Section 5 of the Federal Trade Commission Act.

Given the existence of these prior approvals in specific situations, it appears clear that an industry code directed at the elimination of patently unlawful activities and limited in its scope to appropriate sanctions to correct these abuses will be permissible under the antitrust laws. Of course, great care must be taken in both the drafting of such codes and their enforcement, since arbitrary or discriminatory application of otherwise lawful code provisions can constitute an independent violation of the antitrust laws. Moreover, where membership in the association or industry group is essential to commercial survival of a member firm, imposition of sanctions in the form of liquidated monetary damages is far preferable to the use of dismissal or expulsion as a sanction. In fact, in such cases, dismissal may not be a permissible alternative under the antitrust laws. Once again, good intention alone provides no defense to an antitrust challenge. It is imperative that codes of ethics and their enforcement methodologies receive careful legal review, including submission of proposed codes to the FTC for an advisory opinion in appropriate cases.

III. PROVISIONS FOR SELF-ENFORCEMENT UNDER FEDERAL AND STATE LAWS

While recent legislation has contained only limited grants to nongovernmental entities to regulate themselves, during the Great Depression of the early 1930's Congress enacted a sweeping experiment in industry self-regulation. The National Industrial Recovery Act, promulgated in 1933, authorized the President to approve "codes of fair competition for any trade or industry upon application from groups representing industry members." The only precondition established by this statute was that the President must make the following findings as to the proposed code:

> (1) [T]hat such associations or groups impose no inequitable restrictions on admission to membership therein and are truly representative of such trades or industries or subdivisions thereof, and (2) that such Code or Codes are not designed to promote monopolies or eliminate or oppress small enterprises and will not operate to discriminate against them, and will tend to effectuate the policy of this title: Provided, that such Code or Codes shall not permit monopolies or monopolistic practices. . . . The President may, as a condition of his approval of any such Code, impose such conditions . . . for the protection of consumers, competitors, employees and others, and in furtherance of the public interest, and may provide such exceptions to and exemptions from the provisions of such Code as the President in his discretion deems necessary to effectuate the policy herein declared. [NIRA §3(a), 48 Stat. 196 (1933)]

Once such a "Code of Fair Competition" had been approved by the President, any violation of the code was declared to be "an unfair method of competition in commerce within the meaning of the Federal Trade Commission Act as amended." A number of such codes were quickly developed, approved, and put into effect, including a "Live Poultry Code" applicable to New York City and the metropolitan area surrounding it.

In 1935, this code came before the Supreme Court on an appeal by a New York City slaughterhouse from numerous convictions based upon violations of the code (*A.L.A. Schechter Poultry Corp.* v. *United States*, 295 U.S. 495 [1935]). The Supreme Court first considered the government's argument that the provisions authorizing adoption of codes must be viewed in terms of the grave national crisis which then existed. The Court noted:

> Extraordinary conditions may call for extraordinary remedies. But the argument necessarily stops short of an attempt to justify action which lies outside the sphere of constitutional authority. Extraordinary conditions do not create or enlarge constitutional power. (295 U.S. at 528)

The Court noted further:

> . . . the statutory plan is not simply one for voluntary effort. It does not seek merely to endow voluntary trade or industrial associations or groups with privileges or immunities. It involves the coercive exercise of the law making power. The codes of fair competition which the statute attempts to authorize are codes of laws. (295 U.S. at 529)

The Court summed up the purpose of the NIRA as follows:

> We think the conclusion is inescapable that the authority sought to be conferred by §3 was not merely to deal with "unfair competitive practices" which offend against existing law, and could be the subject of judicial condemnation without further legislation, or to create administrative machinery for the application of established principles of law to particular instances of violation. Rather, the purpose is clearly disclosed to authorize new and controlling prohibitions through codes of laws which would embrace what the formulators would propose, and what the President would approve, or prescribe, as wise and beneficient measures for the government of trades and industries in order to bring about their rehabilitation, correction, and development, according to the general declaration of policy in §1. (295 U.S. at 535)

The Court held the code provisions at issue to be invalid, both on the ground that the attempted delegation of legislative power was unconstitutional and on the further ground that all of the transactions involved occurred in intrastate commerce and had no direct effect on interstate commerce (295 U.S. at 551).

It is important to note that the *Schechter* case does *not* stand for the proposition that self-enforcement of codes of ethics is unconstitutional, but rather that any attempt to delegate legislative authority to establish criminal codes to a private group is not permissible. At one point in its opinion, the Court stated the question thus and gave the following answer:

> . . . [W]ould it be seriously contended that Congress could delegate its legislative authority to trade or industrial associations or groups so as to empower them to enact the laws they deem to be wise and beneficient for the rehabilitation and expansion of their trade or industries? Could trade or industrial associations or groups be constituted legislative bodies for that purpose because such associations or groups are familiar with the problems of their enterprises? . . . The answer is obvious. Such a delegation of legislative power is unknown to our law and is utterly inconsistent with the constitutional prerogatives and duties of Congress. (295 U.S. at 537)

From this perspective, the *Schechter* case leaves open the issue of the extent to which a private organization may be empowered to enforce among its members provisions of substantive law through use of appropriate sanctions.

There are a number of federal regulatory statutes which authorize certain collective activities and provide limited antitrust exemptions for participation in such activities, such as the Interstate Commerce Act's express authorization of industry rate-making activities, subject to subsequent approval by the Interstate Commerce Commission. Even in those instances in which a regulatory statute has bestowed broad authority upon an industry to regulate its own business activity, however, the industry typically remains subject to the general antitrust prohibitions against certain types of anticompetitive behavior. For example, in *Silver* v. *New York Stock Exchange,* 373 U.S. 341 (1963), the Supreme Court held that the power conferred under the Securities Exchange Act of 1934 did not extend to the Exchange's instructions to member firms to deny private wire connections to the non-member plaintiff.

Similarly, the grant under state laws of extensive self-regulatory authority to the medical and legal professions has been held not to exempt a group boycott by doctors from antitrust challenge (*United States* v. *American Medical Association,* 317 U.S. 519 [1943]), and has recently been held not to exempt uniform minimum fee schedules from the prohibitions of the antitrust laws (*Goldfarb* v. *Virginia State Bar*, 421 U.S. 773 [1975]). The particular activities challenged in each of these cases were, however, unequivocally anti-competitive and sought to regulate commercial transactions unrelated to the necessary regulation of the integrity and competence of members of the profession.

Aside from the two practices held unlawful by the Supreme Court, the general legality of the medical and legal professions' self-enforcement of codes for professional and ethical responsibility of their members has never been subject to serious challenge under the antitrust laws. Bearing in mind that the ultimate sanction used by each profession is the barring of unethical or incompetent doctors and lawyers from the practice of their professions, and that in most states, both admission to and expulsion from these professions is largely determined by members of the professions themselves, these are areas in which self-regulation has long been accepted.

IV. CONCLUSION

Although the foregoing discussion has suggested the many potential problems which must be avoided in the drafting of a proper code of conduct, it is equally clear that these legal requirements are neither so complex nor so rigorous as to preclude the development and self-enforcement of effective codes of conduct.

Moreover, given the very real alternative of greater governmental control over particular industries or business practices, an industry or profession confronted by unlawful activities committed by certain of its members can ill afford simply to wait for the government to step in and take over. There is no assurance that legislators considering the problem area will either understand its realities or prescribe appropriate remedies. In any event, the resulting iron law is likely to produce a cure which is far less palatable to the industry or group affected than the disease for which it was prescribed.

16

CODES OF CONDUCT:

Their Sound and Their Fury

ABRAHAM J. BRILOFF

In 1973, as the Watergate labyrinth was just unraveling, Mr. Justice Blackmun lamented that "the glue of our governmental structure seemed to be becoming unstuck." The subsequent disclosures of our corporate Watergating around the globe made it seem that the glue of our economic structure was similarly becoming unstuck. This dual unsticking should create a sense of national urgency because history informs us that when people lose confidence in the leadership of their public and private sectors, that society is ripe for a demagogue.

Let me emphasize at the outset that this malaise is not the fault of my potentially great profession alone; in fact, it may well be that we are not the ones principally responsible for this dilemma. Thus, in the penultimate chapter of my *More Debits Than Credits,*[1] I describe the multiplicity of groups in our total society who are alongside us on this *via dolorosa,* including corporate managements, lawyers, the courts, financial analysts, stock exchanges and their member firms, investors (both institutional and individual), the banks and bankers, and the regulatory agencies of our government (the SEC included) —all of these groups and others can be seen to be *in pari delicto.*

Nonetheless, I will address the issue related to codes of conduct from the vantage point of my profession, undoubtedly because, like Archimedes, I need a place to stand to move the world.

Scan the 1,200 pages (over 700 devoted to the Northrop Corporation) of the single volume representing Part 12 of the Hearings on the Multinational Corporations and United States Foreign Policy[2]—of Senator Church's Subcommittee—to see what happens when the power concentrated in our major corporations becomes so absolute as to corrupt absolutely. More briefly, study the mere 300 pages of the John J.

McCloy Committee Report on the Gulf Oil Corporation[3] to see how power exercised insidiously corrupts insidiously.

Our corporate society represents trillions of dollars in material resources. But going far beyond these material property concentrations, these entities control enormous pools of human resources whose conduct will affect, for good or ill, the universal environment from here to eternity.

The power concentrations are so enormous that we must understand them as more than CPAs' debits and credits, more than earnings per share, more than as price/earnings quotients. We must come to grips with the nature and conduct of our major public corporations with the same degree of circumspection as we would a government. Many of these corporations frequently exercise power beyond that of most governments. And where a particular corporation permits itself to become an extension or agent of a government, then that involvement too should be the subject of full visibility and accountability.

Thus, visibility and accountability are complementary to eternal vigilance—the precondition to our economic and political freedom. We must assure these qualities to both the ostensibly private and the presumptively public sectors of the corporations and governments, respectively.

When the Northrop and Lockheed scandals first surfaced in the hearings before the committees chaired by Senators Church and Proxmire, we were assured that in time we would be "getting a better perspective." We were told that this nation cannot export its higher moral code; bribes may be illegal here, but in the corrupt foreign areas of the world, such payments are completely acceptable.

Indeed, we have gotten a better perspective. With the passage of time we have learned that many nations, contrary to the prevailing myth, take the situation even more seriously than we do. Read the box score on the Japanese indictments, notice that the probe was pursued without let or hindrance, and then decided which nation looked with greater disfavor on corporate corruption. Shame on the judge of the United States District Court who, as a precondition of releasing testimony to the Japanese government, bargained for the immunity of Americans who had made the payments. Shame to the Northrop Corporation shareholders who, at their 1976 annual meeting, warmly applauded the company's chairman, Thomas Jones, a convicted felon. This ovation was given notwithstanding his company's pervasive involvement in illegal and illicit acts at home and abroad. All that seemed to matter to the stockholders was that sales and profits were up, and the stock price was ebullient. Our nation cannot retain its glory if it is obsessed with the pursuit of Mammon to the denigration of our ideals and standards.

According to a *New York Times* feature article in late 1976, Mr. Jones is pictured in imperial garb ready to take Northrop to greater heights. Clearly, we have forgotten the lesson of Watergate (assuming that we had once learned it). Remember Santayana's foreboding that a people which ignores its history is destined to repeat its mistakes.

Turning to the title theme, codes of conduct—those which are of special interest to the accounting profession—I am not especially enthralled by their effectiveness in assuring that my profession fulfill its responsibilities to the total society which, reciprocally, rewards us so abundantly in material and psychic terms.

In the pursuit of our professional calling, we accountants are governed by three separate but related codes which go by the acronyms GAAP, GAAS and COPE—standing, respectively, for "Generally Accepted Accounting Principles," "Generally Accepted Auditing Standards," and "Code of Professional Ethics."

GAAP is supposed to provide us with the theoretical and conceptual infrastructure. GAAS, on the other hand, is supposed to assure the readers of the financial statements to which we add our *imprimatur* that those who undertook the auditing responsibility were of professional caliber, that they went about their tasks with the responsibility fitting to the task, and that the resultant product is entirely reliable. COPE to an important degree subsumes the other two codes but then seeks to impose an overarching discipline by the profession's established professional organization, the American Institute of Certified Public Accountants, over its members. Specifically, COPE addresses itself to the themes of the members' independence, integrity, objectivity, competency, and responsibilities to clients and colleagues.

Such a trilogy of codes should assure the ultimate commitment by my colleagues and a corresponding quality product, resultant audits that provide the public with the full and fair objective history of the corporate enterprise. Such a product would ensure that corporate power is exercised with full visibility and accountability—conditions precedent to the power being utilized in the public interest, rather than for the limited, self-serving objectives of a particular segment thereof.

Why, then, do I have misgivings that with all of GAAP, GAAS, and COPE, my profession is not fulfilling its mandate, to which we are pretending and for which we are paid so handsomely? Principally it is because the trilogy in practice has not produced sweetness and light, nor the true, the good, and the beautiful. And as I see it, the failure derives, in important measure, from the inexorable fact that these codes are the writings of persons who have impacted into them their own particular vested interests and those of their clients. And probably of transcendent import, these codes are interpreted and implemented by an accounting establishment which has its set of vested interests. These interests, in turn, are not necessarily consistent with the social objectives and all too often become instruments for the self-aggrandizement of this accounting establishment and its clients, especially the major corporations, where the economic, political, and social powers are especially concentrated.

Directing our attention to the first of the trinity of codes, that of GAAP, we see the accounting profession groping for its impossible dream, that of developing a codified statement of our underlying body of knowledge. We should be mindful of the diverse forces which have impacted on this body of knowledge. We should know of the ways in which GAAP has become something of a polyglot, reflecting the dictates of economics and the law, as well as of our own profession's evolution. We know the evolution of GAAP reflects the judgments of the courts, the Internal Revenue Service, the Securities and Exchange Commission and other regulatory agencies, and the Congress. It is from all of these forces that our profession has sought some form of logical codification, however tenuous it might be.

In this search for ultimate truths in accountancy, we have had a number of great and noble experiments, each launched with great hope and promise. First, we had the tortured history of CAP, the Committee on Accounting Procedure, followed by that of the APB, the Accounting Principles Board, in their respective endeavors to define the chimera of GAAP. Each body, in its turn, came apart when it proved impotent in the face of the vested interests in corporations and their allies among the accounting firms and in government.

The Financial Accounting Standards Board (FASB) was designed to be the *deus ex machina* to change all this. The high level of compensation of that septenary and its isolation on the high ridges of Stamford, freed from the temptations and tensions of New York City (or some other Sodom), were expected to assure to the Board the intellectual independence required to ferret out divine truth. Alas, early in 1976 we heard from the chairman of that exalted Board that he and his colleagues are not yet in Elysium. Thus, in mid-January, Marshall Armstrong complained to the Securities Regulation Institute that:

> As we solicit views through our discussion memoranda and exposure drafts, we hear from many. A few, like the large public accounting firms and some business enterprises, are, for the most part, interested in achieving standards

> that will eliminate alternative practices, standards that will find their support in the broad, conceptual bases of accounting. Their responses reflect this, and while their views are diverse, their reasoning is generally quite lucid. And others—many others—respond from a vested interest in diversity of practice, from a firm commitment to flexibility of measurement.[4]

Since then, Mr. Armstrong has had further misgivings regarding the Board's prognosis, possibly because he has since heard the emotional, sometimes hysterical, outbursts from the staid bankers who were concerned lest FASB's ruminations might compel them to write down the carrying values of their rotten Real Estate Investment Trust (REIT) loans and their loans to certain third-world countries.

The Board had touched off a great debate by publishing a mere discussion memorandum on the alternative field rules for the accounting by creditors (especially) for so-called "restructured debt"—that which, while not necessarily rotten, is nevertheless seriously overripe. Should a bank which loaned $10 million, for example, to a REIT at 12 percent interest and is now getting no interest, but is hoping that some day portions of the principal will be forthcoming, be permitted to carry the loan on its books at $10 million? Or should it show the marked-down present value of the potential cash flow? Consider the bank which owned $10 million in New York City notes and swapped them in late 1975 for $10 million in MAC bonds, then worth about $7 million; should the swap be deemed an accounting event so as to require the immediate recognition of a $3 million loss? Or should the accountings be made oblivious of the deterioration of value? These questions triggered 800 letters from emotional bankers and some from persons—for example, Dr. Arthur Burns—who were heretofore presumed to be devoid of emotion.

What kind of humbug were the bankers trying to put across? Did they believe that by carrying the REIT or New York City loans, they acquire a $10 million value if they are carried on the bank's books at this amount? It sounds more like a scene from *The Mikado* than from a formal committee formed to discern economic reality and thereupon disclose it to the outside world.

The bankers themselves would undoubtedly tell a potential borrower to stop fantasizing if his books were to value an inventory, or a receivable, at some vestigial historical cost—when it was recognized that the then current value was appreciably less. Just ask these banks how they reported their "restructured debt" instruments on 1975 tax returns. Did they not show, as an immediate charge-off, the $3 million loss on the MAC swap hypothesized above?

If, as I suspect, the Financial Accounting Standards Board is intimidated by this external pressure (as it and its predecessors have been consistently heretofore), the days of the FASB will be numbered—as well they should be.[5]

So much for the principle-determination process. What happens when GAAP is made operational? At the outset, let me state a basic premise to which I subscribe: The underlying theory of accountancy, this GAAP, is tolerably good—if only it were applied in a principled fashion. By way of example, I turn to the Lockheed Syndrome—with emphasis on the "syn"—and the company's accounting for over a half billion dollars of cost overruns incurred on its L-1011, TriStar program. In *More Debits Than Credits,* I accused the corporation and its auditors of having perpetrated a hoax in the very deferral of this grotesque amount. By reference to reports from the General Accounting Office and to generally accepted accounting principles, as well as to self-evident economic truths, I demonstrated that this "asset" was plugged into the balance sheet to maintain some semblance of solvency—however specious—and carried in inventories as a "current asset."

Mirabile dictu! According to its 1975 report, the corporation has now determined to "bite the bullet" (to use a military-industrial metaphor). Did it do so with a *mea culpa?* Not so that we would notice. Instead, we were told by Lockheed that it intended to "take a more conservative stance." And, even then, does this newly discovered conservatism demand that the firm remove the half billion from the balance sheet—as should be done when an asset is demonstrated to be worthless (or even of negative potential utility)? Again to the contrary. The write-off (rip-off) is at the rate of $50 million per year for the decade beginning October 1, 1975. I maintain that this is an abortion of an abomination by attenuation. And as a further sop to "fairness," the remaining debit was stepped down from current assets to the "deferred charges" category, where it remains as an asset nevertheless. Only a company as arrogant as Lockheed would even attempt such a practice; and only an accounting firm securely ensconced within the accounting establishment would associate its name (however tenuously) with such financial statements without being censured by the Securities and Exchange Commission and ostracized by its professional colleagues.

With further reference to GAAP in action, I will comment briefly on the sludge dredged up in the complaint filed in 1976 in the Federal court by Arthur Andersen and Company against certain actions by the Securities and Exchange Commission. What Arthur Andersen seems to find objectionable is the Commission's added requirement that the independent auditor state affirmatively, when his client makes a significant change from one permissible accounting alternative to another, that the change was to a *preferable* one. One might have thought that this requirement would be especially pleasing to Arthur Andersen. After all, that firm more than others has in its public pronouncements beaten the bushes for a transcendent standard of "fairness." That is all that the SEC has suggested by this ASR 177 requirement. But these pronouncements notwithstanding, Arthur Andersen would like to continue to run with the hares and hunt with the hounds. It would like to concur overtly in Texaco's switch from full costing to successful effort accounting, while perpetuating its concurrence in the continued full costing practices at Occidental Petroleum.

As to the GAAS, the second of the codes of conduct governing the profession of accountancy, one might first assert that there is little in that codification with which to disagree. Once again, the evil lies in the practices by which GAAS has been implemented. A fair and just application of standards *qua* standards of auditing would have aborted the contrived transactions involving realty, franchising, and equipment leasing. If the standards had been applied as standards, there would not have been the erosion of the business combinations precepts. There would never have been the fictitious ebullience manifested by the financial reports of our REIT's, land developers, and bank holding companies. If our colleagues had, in fact, recognized the significance of the term "standards" in the phrase "auditing standards," there would never have been the proliferation of litigation against the leaders of our profession for their dismal performance of the attest function.

I am not here alluding to those relatively infrequent cases where our colleagues were sandbagged or snowed by a well-conceived and equally well-concealed fraud. Instead, I am inveighing against those situations in which the audit partners recognized that they were taking what they thought were calculated risks in choosing evil. The unveiling of the sagas regarding slush funds, kickbacks, bribes and similar illicit and illegal payoffs invariably raises several questions: Where were the auditors? Where should they have been? When did they first discern the cancer of corruption (to use Senator Frank Church's metaphor)? Why did they not blow the whistle?

I am aggrieved that in all this, the accountants' established leadership has been so determined to cover their flanks (or whatever part of their anatomy they are covering) that they have not responded to the ultimately critical question. What responsibility will the auditor assume now and henceforth? The profession's credibility with the public is so limited as to demand an immediate, urgent, and unequivocal response.

So it is that I have remonstrated against the AICPA's mountains (those at the AICPA's Auditing Standards Executive Committee, "AuSEC") for their having labored so long and arduously to bring forth the mouse of SAS-5,[6] intended to define the meaning of the critical phrase "present fairly" in our standard form of certificate. On the one hand, such a clear and compelling statement is essential to define more particularly the real responsibility assumed by the independent auditor to third parties when he is presumed to have performed the attest function. On the other hand, such a clear and compelling statement is essential to exorcise the prevailing myth among the masses that it is the independent auditor who has fought the facts and figures and has come forth with a nexus of statements presumed to be the auditor's independent, objective interpretation.

We in the profession know that the realities diverge drastically from the prevalent myth. We know full well that the narrative is that of management and not the auditor's. In short, the phrase "present fairly in accordance with GAAP" is a chimera capable of being defined according to one's tastes, level of sophistication, and the prevailing circumstances. The leadership of the profession endeavors to perpetuate this amorphous condition, since it permits the auditor to earn a professional income while avoiding professional responsibility. The auditor is now capable of charging fees as though he were determined to present a statement which is as fair as he could judge the fair. Yet, in litigation, he can dodge his presumed responsibility by asserting that statements which are "fair in accordance with GAAP" are not necessarily fair—and assert that anyone who presumed otherwise is naïve, or a fool, or both, and should not be entrusted with the interpretation of such a sophisticated tool as a set of financial statements.

More recently, AuSEC heaved and hawed just to bring forth a standard regarding the auditor's responsibility for discerning fraud and other corrupt practices in the client's activities. Basic to these summit deliberations were the public's outcries as to where the auditors *were* and where they *should* have been when all the nefarious conduct involving bribes and political payoffs was being pursued.

Once again, the mountains came forth with a lot of rhetoric full of exegetic analysis but absent a full and fair response to the critical challenge. Once again, the words add up to "we can't." In my view, the profession's abdication of responsibility in these critical areas takes on added significance in the light of the reversal by the Supreme Court of the United States in the *Hochfelder* v. *Ernst & Ernst* case.[7] The Court ruled that an auditor may not be held liable in civil actions by third parties suing under Rule 10(b)-5 of the Securities Act of 1934, despite proven negligence, unless it can be demonstrated that the auditor was conspiratorially or otherwise actively and directly involved in the development of the false or misleading statements. In short, it would appear that the plaintiffs must now be constrained to develop something of a "devil theory" regarding the auditor-defendant if the action for damages is to prevail.

I contend that the abdication of forthright responsibility by the accounting profession, coupled with the abrogation of liability by the courts for an accountant's irresponsibility, means that the archaic standard of *caveat emptor* has been resurrected in the public securities market. When the public discovers this perverse condition, the auditor might well be constrained to associate a skull and bones logo with his signature as a Certified Public Accountant. Thus, the *Hochfelder* exculpation of the accountants will trigger the inculpation of our profession.

We come, then, to the third of the codes which serve to govern the conduct of those engaged in the profession of accountancy, our Code of Professional Ethics, hence our COPE.* This code, like codes of professional ethics generally, is intended as a consciousness-raising book of rules. It is a codification presumed to require that those who pursue the particular calling will be committed to a standard of conduct which is higher than that required generally of the vulgar masses (and no pejorative is intended by the use of the phrase "vulgar masses"). The profession is then presumed to administer this special code of conduct as an incident to its self-regulatory responsibility, i.e., to assure a level of performance higher than that which might be compelled by a court of law. In this very connection, then, I have made it one of my responsibilities to monitor the AICPA's disciplinary practices. The Institute does have a highly structured ethics apparatus, including a Professional Ethics Executive Committee and its subcommittees, the Professional Ethics Division, with a staff, Trial Boards, and sub-boards.

What has this elaborate disciplinary machinery achieved? Over the 90-month period from January 1970 through June 1977, its box score of actions categorized by offense is as follows:

For failure to file (or for false filing) of personal returns	22
Because of revocation or suspension of member's certificate by his state licensing body (where specific reason not indicated)	21
For conviction of bribery	21

* The Rules of Conduct of this Code are reprinted as an Appendix to this essay. See pages 280–87.

For conviction of grand larceny, embezzlement, extortion, theft, perjury, and corresponding high crimes	12
For conviction of mail fraud	8
For failing to disclose (or for false disclosure) to the SEC or IRS	6
For conviction for involvement in the Equity Funding fraud	5
For lack of independence	4
For moral turpitude and other undisclosed crimes	4
For solicitation and advertising	4
For violation of securities laws	2
For refusal to cooperate in grand jury or state investigations	2
For substandard auditing	2
For fraud on CPA exams	1
For failure to pay for securities	1
For obstructing justice	1
For failure to acquire sufficient information	1
For threatening to inform on client	1
For filing false reports with HUD	1
For inadequate disclosure in footnotes	1
For lack of independence in audit	1
A total of	121

These are all grievous offences, and grievously have our colleagues been made to pay for their transgressions. Yet, when I review these actions of the impressive disciplinary apparatus, I cannot help but conclude that the disciplinarians are very much like the blind guides of Scripture, those who strain at the gnat while swallowing camels. Except for the convicted felon in the National Student Marketing fiasco and those involved in Equity Funding, the 121 cases have not addressed the horror stories which have induced the prevailing credibility gap—the cases which I call the Roll of Dishonor, running the alphabetical gamut from Ampex to Yale Express and involving all of the Big Eight accounting firms.

When I expressed a corresponding obloquy in late 1975, the then AICPA chairman, Ivan Bull, took umbrage, pointing to the increased activity on the part of the Institute's disciplinary apparatus. It is true,

Ivan Bull has me by the numbers—of the 121 cases, no fewer than 20 were in the first nine months of 1976 (i.e., the period preceding the initial presentation of this essay). Let us study those 20 reported cases to see how "diligently" the panoply of ethics division, committees, boards, *et al.* pursued its presumptive responsibility of exorcising evil from our profession.

The initial issue of the 1976 *CPA Letter* took care of five of the twenty reported cases. It told of the Equity Funding defendants who pleaded guilty and their two colleagues who were convicted after trial. The former were expelled, the latter suspended pending appeals. *Sic transit gloria mundi!* The issue of the following fortnight told of a member who was suspended for a period of time because he had been involved in the bribing of revenue agents. He had been granted immunity from prosecution and thereby avoided standing trial and possible conviction. That same issue told of another Institute member who had committed the cardinal sin of "soliciting engagements by letter and had conducted a practice under a fictitious name." He was sentenced to "take a course in professional ethics." Ironically, that course was probably written (possibly even taught) by a partner in one of the very firms to which I had referred as having produced the prevailing credibility gap and corporate accountability crisis.

We heard nothing from the Institute's ethics apparatus for the succeeding five months. Then we were advised that three members were expelled—two because of conviction for filing false returns, the third because his state had revoked his license. In August, the wheels of the gods ground out six judgments: one for embezzlement, three for filing false returns, and two for mail fraud. Finally, the ninth month of this test period produced two more expulsions and two suspensions. The former were for grand larceny and perjury convictions; the latter to correspond with the revocation of the CPA licenses by the members' state boards.

That adds up to the twenty. As I reflect on these awesome judgments, I cannot help but conclude that except for the poor unnamed member who did not know how to solicit professionally (the way our prestigious colleagues do), there was really no compelling need for committees or boards, to say nothing of a whole division. All that was needed was an efficient mail clerk to receive advices from various jurisdictions regarding convictions and thereupon to issue the appropriate decree. True, absent this awesome apparatus the poor soliciting scoundrel would not have been apprehended and punished. By my standards, I am prepared to accept that lost opportunity in order to avoid the pretense that we are, in fact, administering a code of ethics worthy of its name.

I have omitted one most dramatic action by the Institute's Trial Board. The initial 1976 CPA letter reported:

> AICPA member Maurice H. Stans, former Secretary of Commerce, has been found by a sub-board of the Trial Board of the AICPA to be not guilty of charges brought by the AICPA's division of professional ethics.
>
> The ethics division's charges arose from Mr. Stans' plea of guilty in Federal Court in Washington in March of last year to five misdemeanors relating to his conduct as Chairman of the Finance Committee to Re-elect the President. It was alleged that the subsequent conviction tended to bring discredit to the profession. Mr. Stans contended that the offenses were minor and technical, that they had been found by the court to be unwillful and that the transactions upon which the convictions were based had been handled by him in good faith.
>
> Following a full-day hearing, the sub-board, on October 28, 1975, found that the charges of the ethics division had not been proved. Mr. Stans requested publication of this finding and the sub-board has authorized this notice.[8]

I have followed Mr. Stans' *modus operandi* through the Watergate hearings, in *All the President's Men,* and in the general press. I maintain that his plea bargaining of crimes into high misdemeanors notwithstanding, I concur with Thomas F. McBride, Esq., of the Watergate Special Prosecutor's office, who asserted, at the time Mr. Stans pleaded guilty to a series of counts, that

> [Mr. Stans] received the sum of thirty thousand dollars ($30,000) from the Minnesota Mining and Manufacturing Company, which sum . . . came from a corporate source, and that Stans either knowingly accepted or acted in reckless disregard in accepting that corporate contribution.[9]

As we read the sordid tales, the horror stories of those days, we are made painfully aware that Mr. Stans' operations at the Committee for the Re-election of Mr. Nixon had developed an extraordinary system of deniability. Granted, Mr. Stans was able to allege that he knew not whence came the bags of money he dumped into his desk drawer, nor where the money went after it left his safe.

But the Institute's Trial Board has found that Stans' actions did not reflect adversely on our profession. Clearly, I must abide by its judgment. But I am bothered nevertheless because, somehow, I cannot help but believe in my "heart of hearts" that the crisis in corporate accountability was induced more by the Stans aberrations than those of the unnamed bloke who was so stupid as to solicit unprofessionally and had to learn ethics at the feet of the sages at the AICPA.

It is essential that the accounting profession (possibly all professions) end its state of hypocrisy, pretending that it is really implementing a code of ethics worthy of its name. The accounting profession (like all professions) must move forthwith to avoid the two-tier, double standard of justice which prevails presently—the one for the affluent and powerful, and the other for the weak with more modest means.

What specific proposals do I have to make to buttress the accounting profession's disciplinary process, and thereby to improve its moral and ethical posture?

My primary recommendation is to break the stranglehold that the oligopoly of firms within our profession presently holds over the principal professional organizations and their various organs. In this endeavor, and specifically with reference to the disciplinary process, I first turned for guidance to the principal recommendation of the Wheat Commission when it formulated its proposals for restructuring the process for the establishment of accounting principles or standards.

When deliberating on the question as to whether the proposed new board should be part-time or full-time, it came down strongly for the latter, saying:

> The major positive arguments for a full-time board are independence and efficiency. It is the doubts cast on the disinterestedness of a part-time board which trouble its critics the most. They assert that board members having a continuing affiliation with their firms or companies must inevitably find their loyalties divided. It is difficult to assess the seriousness of the risk . . . doubts remain. Partner pressure rather than client pressure may well be a greater danger. . . . At all events, it is perhaps the appearance of nonindependence rather than its reality which poses the more significant problem. If there is a widely held supposition—even an erroneous one—that the present board is too responsive to . . . pressure, then its position is weakened. We see no way of avoiding this difficulty except through a full-time board.
>
> Judged also on grounds of efficiency, the drawbacks of the large, part-time board are formidable.
>
> We therefore recommend a full-time board. . . . Above all, they will have no ties except to the board and thus will be seen to have no private interests which might come between them and the public interest: in these circumstances, their disinterestedness should not be seriously questioned.[10]

To demonstrate that the problems of the accounting profession in this area are not *sui generis*, I find myself much informed by a report promulgated in early 1976 by an *ad hoc* committee of the Association of the Bar of the City of New York to study the Association's grievance procedure.[11] Included in the report are a significant number of the observations and recommendations which I believe to be especially appropriate, *mutatis mutandis,* to the corresponding dilemma of the accounting profession.

Thus, among that committee's conclusions and recommendations are:

> In the past few years the legal community has had its consciousness raised in rather dramatic ways to the issues of professional responsibility. The role of lawyers as public servants, in the courts, and in the legal system generally, is under scrutiny. A signal aspect of this scrutiny is the ability and willingness of the profession to regulate and discipline itself. (P. 1)

This led the committee to observe that the legal community is capable of maintaining self-discipline. Such self-discipline, the committee observed, is a privilege entrusted to the profession and is to be made to operate in the public interest. A privilege so bestowed "cannot survive a widespread and growing public belief that a complaint against a lawyer will be either dismissed . . . delayed interminably or result in the grudging imposition of nominal punishment."

To make the disciplinary process more effective, the committee urged that non-lawyers be given a significant role in the "policy making and . . . supervisory aspects of the procedure." Why this proposal?

> . . . Lay members can bring the perspectives of a broader constituency to the process and make the profession more responsive to the demands and expectations of the society it serves. Lay participation would increase public confidence in the determination of the legal profession to maintain effective discipline and in the integrity of the disciplinary system. Increased public confidence might encourage citizens to bring complaints and disclose misconduct.
>
> The experience of those jurisdictions which have introduced lay participation indicates that both the public and the bar benefit from it. Lay members have proved to be articulate advocates of the disciplinary system. (Pp. 13–14, citation omitted)

Nor should the disciplinary process be conducted *in camera.* The report introduces an observation by Mr. Justice Cardozo, thus:

> Reputation in such a calling is a plant of tender growth, and its bloom, once lost, is not easily restored. The mere summons to appear at such a hearing and make report as to one's conduct, may become a slur and a reproach.

The Justice's wisdom notwithstanding, the committee concluded that "confidentiality . . . seriously weakens the grievance system," and went on:

> Undue confidentiality, we believe, is an important source of public skepticism about the legal profession's ability to discipline itself. While we do not doubt the integrity of the Grievance Committee's closed formal hearings, our confidence is not an adequate substitute for that of the public. It does not build confidence if the legal profession can say only "no comment" when the public wants to know whether notorious attorney misconduct is being dealt with. As recently noted by an eminent jurist:
>
> > The lawyer is entitled to protection from false and slanderous accusations. I am convinced, however, that we lost too much by shrouding in secrecy the formal disciplinary hearings.
>
> Unnecessary confidentiality, moreover, runs counter to the legal community's general bias in favor of conducting public business in public. If the accused attorney is to be treated differently from other persons accused of misconduct, there must be an acceptable justification for the exception. We can find no such justification once there has been a finding of probable cause. (Pp. 16–17, citation omitted)

To insure a delicate balance between the professional person's right to privacy and the public's right to know, the committee proposes that confidentiality be maintained during the investigative stages, to determine whether "there is the probable cause that the complaint is well-founded. . . ." But then:

> . . . confidentiality should end, and the accused attorney should not enjoy extraordinary protection. The next stage, a formal hearing, should be open and a record and determination made upon which a court might later base review. (P. 17, citation omitted)

The Commission on Auditor's Responsibilities of the American Institute of Certified Public Accountants (frequently referred to as the "Cohen Commission," after its late chairman, Manuel F. Cohen) may well have taken careful note of, and good counsel from, at least this aspect of the Bar Association's report. Thus, in its March, 1977, report on its tentative conclusions, the AICPA body observed that "self-regulatory efforts by the procession can be substantially improved." To that end, the Cohen Commission concluded:

> Secrecy should be removed from disciplinary actions and from the penalties imposed. Action on alleged violations of professional ethics should not be deferred pending the outcome of litigation except when the accused demon-

strates that the litigation is directly related to the charges. Action should rarely await the outcome of appeals.

This has been an elaborate discourse on the trinity of codes impacting on the accountant in the pursuit of his profession. I have discussed the merits of GAAP, GAAS, and COPE, pointed up their perceptual virtues and perversity in their implementation. But all three codes and any number of additional codes would be to little avail if the professional person required the pressures of codes imposed from without to guide him to the good, the true, and the beautiful in the fulfillment of his professional commitment.

On this score, I cite the cardinal categorical imperative imposed by Immanuel Kant: "Act only on that maxim through which you can at the same time will that it should become a universal law." Thus, a code of conduct should radiate from within and not merely reflect the dictates of the consensus.

And in this connection, I find myself going back to the concluding segment of my *Unaccountable Accounting*, written, it should be noted, in the relatively bucolic age known as Pre-Watergate. That segment was addressed "To the Alpha and Omega."

> In essence I am seeking to impose a "Nuremberg Code" on each of us engaged in our professional pursuit, a code whereby we commit ourselves to implement a standard of fairness, even though a contrary result could well be subsumed under GAAP. And we should adhere to the code in spite of "superior orders" or clients' directives. This is an awesome burden which I am imposing on the individual—but it is only through the acceptance of just such a burden that accountants will have a fair claim to professional recognition.
>
> In thus returning to the individual accountant and his sense of the fair, the good, and the right, I am repeating the saga of Herman Hesse's literary character, Siddhartha, who set out on a long quest in search of the ultimate answer to the enigma of man's role on earth. After a tortuous road he discovers that the answer was always within the self, and it is there that one is constrained to search for it.[12]

NOTES

1. Abraham J. Briloff, *More Debits Than Credits* (New York: Harper & Row, 1976), Chapter 12, pp. 344–415.
2. U.S. Senate, Subcommittee on Multinational Corporations of the Committee on Foreign Relations, "Multinational Corporations and United States Foreign Policy," Part 12, hearing dates May 16, 1975 *et seq*.

3. Report of the Special Review Committee of the Board of Directors of Gulf Oil Corporation, submitted December 30, 1975, to the U.S. District Court (D.C.) in the Matter of *Securties and Exchange Commission* v. *Gulf Oil Corporation et al.* (Civil Action No. 75-0324).
4. Marshall Armstrong, "The Politics of Establishing Accounting Principles," an address before the Securities Regulation Institute, San Diego, January 17, 1976.
5. New Year's Eve, 1976, brought forth the FASB exposure draft on this restructured debt issue. As one banker observed, "It's a Christmas present one week late." This euphoria resulted from the fact that the prestigious accounting board collapsed in the face of the enormous pressures. The draft was ultimately codified into GAAP as Financial Accounting Statement Number 15 (1977).
6. American Institute of Certified Public Accountants, Statement on Auditing Standards, No. 5, July, 1975, ". . . the Meaning of 'Present Fairly' in Conformity with Generally Accepted Accounting Principles." (Presently incorporated in *AICPA Professional Standards: Auditing,* Section 411.)
7. *Hochfelder* v. *Ernst & Ernst,* 425 U.S. 185 (1976).
8. *The CPA,* January 12, 1967, p. 4.
9. From statement by Thomas F. McBride of the Watergate Special Prosecutor's Office before U.S. District Judge John Lewis Smith, Jr., March 12, 1975 in the matter of *United States* v. *Maurice H. Stans,* Criminal 75-163 (U.S.D.C.-D.C.).
10. *Establishing Financial Accounting Standards,* Report of the Study on Establishment of Accounting Principles (New York: AICPA, 1972).
11. *Report on the Grievance System by the Ad Hoc Committee on Grievance Procedures* (New York: Association of the Bar of the City of New York, January 26, 1976).
12. Abraham J. Briloff, *Unaccountable Accounting* (New York: Harper & Row, 1972), p. 335.

APPENDIX*

RULES OF CONDUCT

In the footnotes below, the references to specific rules or numbered Opinions indicate that revised sections are derived therefrom; where modifications have been made to the present rule or Opinion, it is noted. The reference to "prior rulings" indicates a position previously taken by the ethics division in response to a specific complaint or inquiry, but not previously published. The reference to "new" indicates a recommendation of the Code Restatement Committee not found in the present Code or prior rulings of the ethics division.

Definitions

The following definitions of terminology are applicable wherever such terminology is used in the rules and interpretations.

Client. The person(s) or entity which retains a member or his firm, engaged in the practice of public accounting, for the performance of professional services.

Council. The Council of the American Institute of Certified Public Accountants.

Enterprise. Any person(s) or entity, whether organized for profit or not, for which a CPA provides services.

Firm. A proprietorship, partnership or professional corporation or association engaged in the practice of public accounting, including individual partners or shareholders thereof.

Financial statements. Statements and footnotes related thereto that purport to show financial position which relates to a point in time or changes in financial position which relate to a period of time, and statements which use a cash or other incomplete basis of accounting. Balance sheets, statements of income, statements of retained earnings, statements of changes in financial position and statements of changes in owners' equity are financial statements.

Incidental financial data included in management advisory services reports to support recommendations to a client, and tax returns and supporting schedules do not, for this purpose, constitute financial statements; and the statement, affidavit or signature of preparers required on tax returns neither constitutes an opinion on financial statements nor requires a disclaimer of such opinion.

Institute. The American Institute of Certified Public Accountants.

Interpretations of Rules of Conduct. Pronouncements issued by the Division of Professional Ethics to provide guidelines as to the scope and application of the Rules of Conduct.

Member. A member, associate member or international associate of the American Institute of Certified Public Accountants.

Practice of public accounting. Holding out to be a CPA or public accountant and at the same time performing for a client one or more types of services rendered by public accountants. The term shall not be limited by a more restrictive definition which might be found in the accountancy law under which a member practices.

Professional services. One or more types of services performed in the practice of public accounting.

Applicability of Rules

The Institute's Code of Professional Ethics derives its authority from the bylaws of the Institute which provide that the Trial Board may, after a hearing, admonish, suspend or expel a member who is found guilty of infringing any of the bylaws or any provisions of the Rules of Conduct.[1]

The Rules of Conduct which follow apply to all services performed in the practice of public accounting including tax[2] and management advisory services[3] except (a) where the wording of the rule indicates otherwise and (b) that a member who is practicing outside the United States will not be subject to discipline for departing from any of the rules stated herein so long as his conduct is in accord with the rules of the organized accounting profession in the country in which he is practicing.[4] However, where a member's name is associated with financial statements in such a manner as to imply that he is acting as an independent public accountant and under circumstances that would entitle the reader to assume that United States practices were followed, he must comply with the requirements of Rules 202 and 203.[5]

A member may be held responsible for compliance with the Rules of Conduct by all persons associated with him in the practice of public accounting who are either under his supervision or are his partners or shareholders in the practice.[6]

A member engaged in the practice of public accounting must observe all the Rules of Conduct. A member not engaged in the practice of public accounting must observe only Rules 102 and 501 since all other Rules of Conduct relate solely to the practice of public accounting.[7]

A member shall not permit others to carry out on his behalf, either with or without compensation, acts which, if carried out by the member, would place him in violation of the Rules of Conduct.[8]

[1] Bylaw Section 7.4.
[2] Opinion No. 13.
[3] Opinion No. 14.
[4] Prior ruling.
[5] Rules 2.01, 2.02, 2.03 and prior rulings.
[6] New.
[7] New.
[8] Opinion No. 2.

Independence, integrity and objectivity

Rule 101—Independence. A member or a firm of which he is a partner or shareholder shall not express an opinion on financial statements of an enterprise unless he and his firm are independent with respect to such enterprise.[9] Independence will be considered to be impaired if, for example:

A. During the period of his professional engagement, or at the time of expressing his opinion, he or his firm
   1. Had or was committed to acquire any direct or material indirect financial interest in the enterprise;[10] or
   2. Had any joint closely held business investment with the enterprise or any officer, director or principal stockholder thereof which was material in relation to his or his firm's net worth;[11] or
   3. Had any loan to or from the enterprise or any officer, director or principal stockholder thereof.[12] This latter proscription does not apply to the following loans from a financial institution when made under normal lending procedures, terms and requirements:
      (a) Loans obtained by a member or his firm which are not material in relation to the net worth of such borrower.
      (b) Home mortgages.
      (c) Other secured loans, except loans guaranteed by a member's firm which are otherwise unsecured.[13]

B. During the period covered by the financial statements, during the period of the professional engagement or at the time of expressing an opinion, he or his firm
   1. Was connected with the enterpise as a promoter, underwriter or voting trustee, a director or officer or in any capacity equivalent to that of a member of management or of an employee;[14] or
   2. Was a trustee of any trust or executor or administrator of any estate if such trust or estate had a direct or material indirect financial interest in the enterprise; or was a trustee for any pension or profit-sharing trust of the enterprise.[15]

The above examples are not intended to be all-inclusive.

[9] Rule 1.01 ("shareholder" added to recognize corporate practice).
[10] Rule 1.01.
[11] Prior rulings.
[12] Prior rulings.
[13] Opinion No. 19.
[14] Rule 1.01 (present Rule 1.01 uses the phrase "key employee").
[15] Prior rulings. In order that a member may arrange an orderly transition of his relationship with clients, section B2 of Rule 101 relating to trusteeships and executorships will not become effective until two years following the adoption of these Rules of Conduct.

Rule 102—Integrity and objectivity. A member shall not knowingly misrepresent facts, and when engaged in the practice of public accounting, including the rendering of tax and management advisory services, shall not subordinate his judgment to others.[16] In tax practice, a member may resolve doubt in favor of his client as long as there is reasonable support for his position.[17]

Competence and technical standards

Rule 201—Competence. A member shall not undertake any engagement which he or his firm cannot reasonably expect to complete with professional competence.[18]

Rule 202—Auditing standards. A member shall not permit his name to be associated with financial statements in such a manner as to imply that he is acting as an independent public accountant unless he has complied with the applicable generally accepted auditing standards promulgated by the Institute. Statements on Auditing Procedure issued by the Institute's committee on auditing procedure are, for purposes of this rule, considered to be interpretations of the generally accepted auditing standards, and departures from such statements must be justified by those who do not follow them.[19]

Rule 203—Accounting principles. A member shall not express an opinion that financial statements are presented in conformity with generally accepted accounting principles if such statements contain any departure from an accounting principle promulgated by the body designated by Council to establish such principles which has a material effect on the statements taken as a whole, unless the member can demonstrate that due to unusual circumstances the financial statements would otherwise have been misleading. In such cases his report must describe the departure, the approximate effects thereof, if practicable, and the reasons why compliance with the principle would result in a misleading statement.[20]

Rule 204—Forecasts. A member shall not permit his name to be used in conjunction with any forecast of future transactions in a manner which may lead to the belief that the member vouches for the achievability of the forecast.[21]

[16] New.
[17] Opinion No. 13.
[18] New.
[19] New (replaces Rules 2.01-2.03).
[20] New (replaces Rules 2.01-2.03).
[21] Restatement of Rule 2.04.

Responsibilities to clients

Rule 301—Confidential client information. A member shall not disclose any confidential information obtained in the course of a professional engagement except with the consent of the client.[22]

This rule shall not be construed (a) to relieve a member of his obligation under Rules 202 and 203, (b) to affect in any way his compliance with a validly issued subpoena or summons enforceable by order of a court, (c) to prohibit review of a member's professional practices as a part of voluntary quality review under Institute authorization or (d) to preclude a member from responding to any inquiry made by the ethics division or Trial Board of the Institute, by a duly constituted investigative or disciplinary body of a state CPA society, or under state statutes.[23]

Members of the ethics division and Trial Board of the Institute and professional practice reviewers under Institute authorization shall not disclose any confidential client information which comes to their attention from members in disciplinary proceedings or otherwise in carrying out their official responsibilities. However, this prohibition shall not restrict the exchange of information with an aforementioned duly constituted investigative or disciplinary body.[24]

Rule 302—Contingent fees.[25] Professional services shall not be offered or rendered under an arrangement whereby no fee will be charged unless a specified finding or result is attained, or where the fee is otherwise contingent upon the findings or results of such services. However, a member's fees may vary depending, for example, on the complexity of the service rendered.[26]

Fees are not regarded as being contingent if fixed by courts or other public authorities or, in tax matters, if determined based on the results of judicial proceedings or the findings of governmental agencies.[27]

Responsibilities to colleagues

Rule 401—Encroachment.[28] A member shall not endeavor to provide a person or entity with a professional service which is currently provided by another public accountant except:

1. He may respond to a request for a proposal to render services and

22 Restatement of Rule 1.03.
23 Prior rulings.
24 New.
25 Restatement of Rule 1.04.
26 New.
27 Rule 1.04.
28 Restatement of Rule 5.01.

may furnish service to those who request it.[29] However, if an audit client of another independent public accountant requests a member to provide professional advice on accounting or auditing matters in connection with an expression of opinion on financial statements, the member must first consult with the other accountant to ascertain that the member is aware of all the available relevant facts.[30]

2. Where a member is required to express an opinion on combined or consolidated financial statements which include a subsidiary, branch or other component audited by another independent public accountant, he may insist on auditing any such component which in his judgment is necessary to warrant the expression of his opinion.[31]

A member who receives an engagement for services by referral from another public accountant shall not accept the client's request to extend his service beyond the specific engagement without first notifying the referring accountant, nor shall he seek to obtain any additional engagement from the client.[32]

Rule 402—Offers of employment. A member in public practice shall not make a direct or indirect offer of employment to an employee of another public accountant on his own behalf or that of his client without first informing such accountant. This rule shall not apply if the employee of his own initiative or in response to a public advertisement applies for employment.[33]

Other responsibilities and practices

Rule 501—Acts discreditable. A member shall not commit an act discreditable to the profession.[34]

Rule 502—Solicitation and advertising. A member shall not seek to obtain clients by solicitation.[35] Advertising is a form of solicitation and is prohibited.[36]

Rule 503—Commissions. A member shall not pay a commission to obtain a client, nor shall he accept a commission for a referral to a client of products or services of others.[37] This rule shall not prohibit payments for the purchase of an accounting practice[38] or retirement payments to individuals formerly

[29] Rule 5.01.
[30] New.
[31] Opinion No. 20.
[32] Rule 5.02 restated to include prior rulings.
[33] Rule 5.03, "or that of his client" added.
[34] Rule 1.02.
[35] Rule 3.02.
[36] Rule 3.01.
[37] Restatement of Rule 3.04.
[38] Prior rulings.

engaged in the practice of public accounting or payments to their heirs or estates.[39]

Rule 504—Incompatible occupations. A member who is engaged in the practice of public accounting shall not concurrently engage in any business or occupation which impairs his objectivity in rendering professional services or serves as a feeder to his practice.[40]

Rule 505—Form of practice and name. A member may practice public accounting, whether as an owner or employee, only in the form of a proprietorship, a partnership or a professional corporation whose characteristics conform to resolutions of Council.[41] (See Appendix B, page 28.)

A member shall not practice under a firm name which includes any fictitious name, indicates specialization or is misleading as to the type of organization (proprietorship, partnership or corporation).[42] However, names of one or more past partners or shareholders may be included in the firm name of a successor partnership or corporation.[43] Also, a partner surviving the death or withdrawal of all other partners may continue to practice under the partnership name for up to two years after becoming a sole practitioner.[44]

A firm may not designate itself as "Members of the American Institute of Certified Public Accountants" unless all of its partners or shareholders are members of the Institute.[45]

[39] Opinion No. 6.
[40] Restatement of Rule 4.04.
[41] Rule 4.06.
[42] Prior rulings.
[43] Rule 4.02.
[44] Prior rulings.
[45] Rule 4.01.

VII

RESPONSIBILITY TO FUTURE GENERATIONS

17

FAMINE OR FOOD:

Sacrificing for Future or Present Generations

MICHAEL D. BAYLES

Very few people starve to death. Instead, malnourished and undernourished people, especially infants and children, are highly susceptible to various diseases and frequently die from them. The diets of millions of people, especially in the less developed countries, are nutritionally deficient in proteins, vitamins, and even calories. As the world has a 1.8 percent rate of population growth and a possibly deteriorating climate, some people forecast more widespread famine within the next two decades. Yet, others maintain that sufficient food may be produced to feed a world population ten times its current size and that the problems are in methods of production and distribution.

This situation has resulted in a debate over foreign aid, especially food aid, by developed countries to less developed ones. Some writers maintain that because of government and business policies, including failure to give aid, most citizens of developed countries are responsible for killing people in less developed countries. Other writers claim that no food aid should be given because it would lead to the death and suffering of even more people in future generations. These arguments, frequently stated in terms of lifeboat ethics, often fail to distinguish between the issues of how much aid developed countries should provide, to whom it should be provided, and the form it should take. That is, they do not distinguish between the issues of amount, distribution, and kind of aid. This essay focuses on the issue of amounts of aid, primarily food aid, with a brief concluding discussion of issues of distribution.[1]

LIFEBOAT MORALITIES

The lifeboat analogy is the favorite of those discussing the morality of aid, but it is applied in different ways to arrive at opposite conclusions. Onora Nell and Garrett Hardin, for example, each use it and arrive at widely differing conclusions. Nell has used it to claim that people have a duty to prevent and postpone famine deaths, and that if they do nothing, they "bear some blame for some deaths."[2] She bases her argument on the principle that people have a moral right not to be killed and an analogy between earth and a lifeboat. The earth is currently a well-equipped lifeboat, but sometime in the not too distant future it will be an ill-equipped one. If population continues to grow at current rates, there will not be enough food for everyone. As the earth is currently a well-equipped lifeboat, the only justification for killing people is self-defense. People in developed countries kill some people, who now die of starvation, by adopting policies which lead to their deaths. They have violated the victims' rights not to be killed, and hence, Nell argues, many people in developed countries are responsible for deaths in less developed ones.

Nell's argument for food aid and other policies to avoid deaths in less developed countries is impressive. The basic right not to be killed is widely accepted. However, Nell fails to establish that people in developed countries violate the right to life of people who starve in less developed ones. For one person to violate a second person's right to something, his conduct must (1) be contrary to a correlative duty, (2) cause the second person to be deprived of the thing in question, and (3) exhibit an appropriate mental state. Nell fails to establish each of these points.

1. Rights imply correlative duties of others. If one person has a right of ownership to a car, others have a duty not to use it without his permission. There are exceptions when it may be permissible to use a person's car without his permission—e.g., it is the only means to take a seriously injured person to a hospital. The right to life implies a correlative duty of others not to kill one.

There are three reasons for denying that most people in developed countries act contrary to the correlative duty of the right to life of persons who starve in less developed countries. First, Nell's argument that people kill rather than allow others to die is unsatisfactory. She appears to use the concept of killing so that any death one could have prevented is killing. She claims that conduct resulting in death is killing rather than allowing to die if the death would not have occurred had one had no causal influence or done other acts. However, whenever it makes sense to speak of allowing a person to die, there is something one could do to prevent his death. So, if one had done that other act, he would not have died. Nell may have intended to claim that one kills when one's action causally contributes to a person's death. If so, then she merely requires condition (2) above. Thus, she has not shown that starvation victims are killed rather than merely allowed to die.

Second, Nell recognizes only two conditions in which killing may be permissible: self-defense and unavoidability. However, many people recognize other conditions. Directly relevant is the partial waiver of the right not to be killed by the assumption of risk. If a person knowingly and voluntarily participates in an activity in which he risks being killed, then his right is not violated if he is killed in a way falling within the assumed risks—for example, an auto racing driver killed in a wreck. This consideration is especially pertinent to Nell's argument about foreign investment in less developed countries resulting in deaths. If the governments and people of these countries knowingly and voluntarily participate in or allow such investments, then resulting deaths do not violate rights. Of course, one may argue that the people of these countries do not so participate in the investment activities, but this claim requires a complex factual analysis of the political situation in each country. If the people do not consent but the governments do, then primary responsibility lies with the governments rather than the corporations. In any case, to the extent that people do so participate, deaths within the assumed risk do not violate rights.

Third, Nell assumes that "if A has a right to defend himself against B, then third parties ought to defend A's right."[3] But a duty to defend others from being killed does not follow directly from their right not to be killed. The duty to defend others is not coextensive with their rights. In some legal systems, the justification of defense of others is not as broad as that of self-defense.[4] More significantly, even if the *right* of defense of others is coextensive with that of self-defense, there is no good reason to make a *duty* of defense of others that broad. It would be absurd to impose a duty to defend others even in those circumstances in which they choose not to exercise their rights of self-defense. Consequently, one does not violate a duty to defend a person merely because one fails to prevent someone else from pursuing policies which may entitle the first one to exercise a right of self-defense. Thus, those who do not participate in activities resulting in avoidable deaths in less developed countries are not to blame for them merely because they do not try to prevent the activities.

2. The brunt of Nell's argument is that people in developed countries

are causally responsible for deaths in less developed ones. But this argument is not straightforward. She relies upon the responsibility of stockholders and citizens for corporate and government policies. However, this is role responsibility, that of being ultimately in charge because of one's position. Stockholders and citizens do not cause the deaths any more than an employer causes the negligent automobile accident of an employee many miles away. While there may be good reasons to hold an employer financially liable for such an accident, he did not cause it. Moreover, due to the vast number of stockholders of corporations or citizens of countries, the ability of any one of them to effect a change in policy is quite small. In fact, none can do so acting alone.

3. Nor does Nell establish that the people exhibit an appropriate mental state. The appropriate mental states for moral responsibility (violating a right) are acting intentionally, knowingly, recklessly, or negligently with respect to an occurrence. Responsibility varies, being correspondingly less as the mental state is one of intention through negligence. Citizens, stockholders, and even corporate and government officials usually do not intentionally or knowingly cause the deaths. Corporate or government officials may sometimes act recklessly with respect to the possibility of deaths, but stockholders and citizens are at most negligent in failing to discover corporate or government policies and their consequences. It is highly doubtful that they have any duty to determine the details of those policies and whether they in fact result in higher or lower death rates in less developed countries than alternative policies would.

A lifeboat analogy has been used by Garrett Hardin to argue for a conclusion opposite to that of Nell—namely, that developed countries should not provide food aid to less developed ones.[5] Hardin treats each country as a separate lifeboat. The developed countries constitute well-equipped lifeboats, while the less developed countries are ill-equipped ones. Each boat has a maximum carrying capacity, and there are people swimming in the water around the lifeboats. The issue, as Hardin poses it, is what the people in the well-equipped lifeboats ought to do with respect to those who ask either to be let into their boats or to be given some of their provisions.

Suppose a well-equipped lifeboat has a capacity of 60 people and only 50 on board, which provides a safety factor. There are three options available to the people on it. (1) They may take in all who ask, but then the boat will become overcrowded and sink. (2) They may admit only 10 persons, which fills the boat to capacity, eliminates the safety factor, and requires the difficult choice of the 10 to be admitted.

(3) They may not admit anyone. If some are troubled by their good luck in being on a well-equipped boat, they may surrender their place to another, and soon the boat will be occupied only by those whose consciences are not so troubled. As the example is stated, it primarily applies to immigration policy. However, a similar argument applies with respect to giving provisions to the ill-equipped boats: the provisions may be shared equally, only the "extra" provisions may be given, or none may be given.

Hardin contends that giving aid will merely postpone the inevitable famine and make it worse than it would otherwise be. In the absence of external aid, population growth follows an irregular curve around the carrying capacity of a country. When the population becomes too large for the food supply, natural factors, such as disease and starvation, reduce the population so that it is within the carrying capacity. However, if, as the population approaches its limits, aid is provided to avoid emergencies, then the population stabilizes at a larger number. It will then expand to an even greater size, creating another emergency. If aid is again provided, the cycle is repeated. Hardin calls this "the ratchet effect." Sooner or later, the problem will become too large for outside sources to provide sufficient aid to prevent famine. Since there will then be many more people than in the beginning, there will be more deaths due to starvation than there would have been had no aid been provided initially.

Finally, Hardin notes that a World Food Bank operates as such an aid program and not as a normal bank. In a normal bank, each person makes deposits and is allowed to withdraw only from what he has deposited. (Loans do not really affect this point, because a borrower must return what he has borrowed, plus interest.) In a World Food Bank, however, food-short countries will not make deposits, only withdrawals, or at least, their net deposits will be less than their net withdrawals. Countries with food surpluses, however, will make larger deposits than withdrawals. Consequently, instead of operating on a regular banking principle, a World Food Bank is a method of making transfers from the food-rich to the food-poor. Thus, it effectively operates as aid in emergency situations and produces the results considered above.

Hardin's argument has been criticized for not giving sufficient weight to the lives of those currently alive. As Daniel Callahan notes, even granting that one has a greater obligation to one's grandchildren than to others due to one's responsibility for their birth, Hardin's proposal still involves a deliberate decision to allow people to die who might otherwise survive a while.[6] In particular, it sacrifices those currently alive for the sake of the unborn—the greater number who would die later and the well-being of future generations in the developed countries. Hardin asserts, for example, "Every life saved this year in a poor country diminishes the quality of life for subsequent generations," and "To be generous with one's own possessions is one thing; to be generous with posterity's is quite another."[7] An ecologically oriented ethics, he also claims, "insists that the needs of posterity be given a weighting commensurate with those of the present generation. The economic prejudice that leads to a heavy discounting of the future must be balanced by a recognition that the population of posterity vastly exceeds the population of the living."[8] On the other side, Callahan claims, "While we surely have obligations to future generations, our more immediate obligation is toward those now alive. There is no moral justification for making them the fodder for a higher quality of life of those yet to be born, or even for the maintenance of the present quality of life."[9]

In effect, then, food aid raises the dual issues of sacrifice within a generation and between generations. By "sacrifice" is meant people not having some goods or benefits they would have were a different policy adopted. As just explained, not giving aid in order to secure the quality of life of future people involves sacrifice between generations, because people currently alive will die sooner than they would were aid given. The giving of aid involves the sacrifice of those in the future for the sake of those presently alive. It also involves sacrifice within a generation, because the better-off give up some goods they might have for the sake of the worse-off. Thus, there are two sacrifices involved in the provision of aid. An argument in favor of it will have to take both into account.

OBLIGATIONS TO FUTURE GENERATIONS

The crux of Hardin's argument is that if food aid is given now, more people will starve later, and that consequences in future generations must be weighed equally with famine deaths which would occur now were food aid withheld. However, on current ethical theories, it is not obvious that future famines would be a great evil. In support of Hardin's thesis, Joseph Fletcher and Tristram Engelhardt have both made the seemingly plausible claim that future famines would injure those alive then. Fletcher contends that those who subsequently starve will be injured.[10] However, one must compare what happens on the two alternatives of aid and no aid. If aid is given, they will live for a while and then starve. If aid is not given, many of them will never live at all. For the first option to be worse for them than the second, their brief lives would have to be so miserable that it would be better for them not to live at all; that is, their lives would have to be not worth living. While Fletcher holds that a rather high quality of life is essential for life to be worth living, there is little if any reason to believe that these people would not attain that level.[11]

Engelhardt presents a more subtle analysis. He does not claim that those who starve but would not have been born had no aid been given are injured. Instead, he claims the injury is only to those who would have been born anyway, their quality of life being lower than it would otherwise have been.[12] That is, he claims that if aid is not given, there will be a future generation of i people. If aid is given, then there will be a future generation of $i + j$ people. Fletcher claims the injury is to members of the i or j classes who starve. Engelhardt claims it is only to members of the i class who are worse off, whether by starvation or otherwise, due to the existence of the members of j.

Engelhardt's analysis will not do. First, it assumes that the i class will be composed of the same individuals on either alternative, so that one may speak of the same persons being better off on one than the other. But the i class will not consist of the identical individuals on the two alternatives. The difference in policies would change the timing of births and different people would be born, at least as different as siblings.[13] Second, even if they were the same individuals, they would not be injured in the sense of having had goods taken from them. Instead, they would merely fail to receive benefits they might otherwise have had. The failure to provide benefits is not morally equivalent to injuring people. It is morally one thing not to give an employee a Christmas bonus and quite another to reduce his salary.[14]

Instead, the issue should be formulated as follows: Assuming justice within each generation, how much quality of life ought one generation to sacrifice for another? "Quality of life" does not refer simply to material goods, but to the total natural and social environment providing conditions and opportunities for a meaningful and rewarding life. A sacrifice may be by present generations for future ones, or by future generations for present ones. These sacrifices will usually be of forgone benefits rather than incurred losses. For example, one argument for Stalinist repression in the Soviet Union during the 1930's was that drastic measures were needed then to bring about rapid industrialization and improve the quality of life for future generations. Thus, the generation of the 1930's was sacrificing for the generation of the 1960's. On the opposite side, some people believe that the high quality of life in the United States during the 1950's and 1960's consumed material resources and created such pollution that subsequent generations—perhaps those of the 1970's and 1980's—will have a lower quality of life, at least as measured by pollution, resource consumption, and prices than they would otherwise have had. Such sacrifices by future generations are always an involuntary forgoing of benefits.

It is a well-accepted principle of moral philosophy that time is irrelevant.[15] Whether an event or state of affairs occurs at one time or another is irrelevant to the evaluation of it or the choice that produces it. Of course, events in the far future may be less certain to occur than those in the near future and, therefore, less heavily weighted in evaluating the consequences of choices, but the lesser weighting is due to the uncertainty, not the time per se. Consequently, if one seeks to obtain the highest average or total quality of life over time, one may prefer a lower than possible quality of life now because it will be outweighed by a much higher one later, or vice versa. A high rate of savings and investment now would provide a lower quality of life at present than a low rate, but it might pay off in a much higher quality of life in the future. It follows on these principles that "generational sacrifice," the sacrifice of the well-being of one generation for another, is justifiable provided less suffering or greater well-being results when the well-being of all those existing at any time is considered.

The problem is, what precisely is the obligation of the present generation to future ones? One may perhaps approach it by roughly ranking the degrees of concern or benevolence one might have for future generations. A minimum concern might be that future generations survive long enough to reproduce themselves and continue the species. This degree of concern is roughly comparable to one that currently existing children survive to adulthood. A greater degree of concern would be to ensure that future generations have a minimum level of well-being or quality of life—higher than that for mere survival but not as high as it might be. Such a concern is analogous to a concern to provide a floor of welfare for currently existing people. A concern for a minimum well-being does not involve a concern for others equal to that for oneself, for presumably, one desires more than minimum well-being for oneself. Consequently, a third level of concern would be that future generations be as well-off as the present. One would want others to have as much as one does oneself. Finally, one could be altruistic in the sense of wanting others to be better off than oneself. Thus, in order to make life even better for future generations, one would be willing to take less than what would leave them equally well-off.

Principles of moral obligation to future generations embody one of these levels of concern. The issue is what principle may be justified.

As a first step in determining what principle is justified, one may explore the implications of there being any obligation to future generations.

Any degree of concern for future generations will impose some moral constraints upon what the present generation may do. Suppose the present generation has a quality of life of level *p*. Further, suppose it is accepted that future generations ought to be able to have a minimum quality of life of level *q*, which is less than *p*. This obligation may be called the minimum principle. Then the present generation is not justified in acting in ways which would prevent future generations from attaining level *q*. However, this restriction leaves open the size of future generations, although freedom to have the children one desires is a factor in determining present quality of life. The quality of life of the present generation may not prevent the next generation from attaining level *q*, but it may necessitate its being smaller than it would otherwise have had to be. The present generation thus will not be able to have as many children as it might otherwise, but one may assume other benefits more than make up for losses in not having children. Consequently, provided each subsequent generation is smaller than it otherwise need to have been, it is possible for each generation to live at a higher quality of life (involving a more rapid consumption of resources, etc.). In short, the size of future generations may be sacrificed for present quality of life.

However, there is a limit to such sacrifices. If future generations become too small, they will not be able to sustain a quality of life of level *q*. Moreover, since moral principles are indifferent to time, it makes no difference whether it is a generation in the near or distant future which would become too small to sustain a population with a quality of life at level *q*. Consequently, the present generation is constrained to act so that a population with a quality of life of level *q* is indefinitely sustainable. The population which may be sustainable at that level might be considerably smaller than the present one. Moreover, the present generation may enjoy a much higher quality of life than *q*, and future generations may be able to live only at level *q*. Whether that is possible largely depends on how high level *q* is. In any case, this argument establishes a basic element of an obligation to future generations. Any obligation to future generations, even for mere survival to adulthood, requires the present generation to act so that future generations can be indefinitely sustained at that quality of life.

Suppose generations live in ways which are not indefinitely sustainable—for example, by producing pollution, consuming resources, or simply having large numbers of children. Then sooner or later, there will be a generation which cannot live above level q and enable the following generations to attain a quality of life of level q. Consequently, in the long run, whatever level q is, if previous generations do not provide for more than that level, the population will be unable to live above it and indefinitely sustain a population at level q. Thus, the ultimate principle must be that a generation is not justified in acting in ways which will not permit the continuance of a population at the current quality of life. In short, the only principle of obligation to future generations which can be adopted in the long run is that the present generation act so that a population is indefinitely sustainable at the present quality of life.

While this is the only principle which may be justified in the long run, it does not follow that it must be adopted now. The conduct of the present generation may not allow for future ones to have an equally high quality of life but only a minimum level. However, one cannot let the standard of a minimum level to be sustained vary over time. If the present generation is at level $q + 2$, it might claim that the minimum level owed to future generations is q and so provide for the next generation. But the next generation may claim that the minimum level owed is $q - 1$ and so live at level $q + 1$. Repeated resetting of the minimum level owed would result in a constantly diminishing quality of life. Hence, the standard of the level owed future generations must be constant.

If one thinks q is the minimum owed others, one should act to assure its attainment. Only if people in future generations will agree to the principle can one be assured that it will be adhered to and no resetting occur. Thus, one should adopt a principle to which future generations may also be expected to agree and adhere. The principle states what generations are obligated to provide future ones, what it would be wrong not to provide. It thus sets limits to their self-seeking. Consequently, one may assume each generation will seek as high a quality of life for itself as is morally permissible. Each generation will hope previous ones set q as high as possible. The result, then, is that all generations will agree only if q is as high a quality of life as may be indefinitely sustainable. Consequently, the only principle which may realistically be expected to be adopted and adhered to is that each generation is obligated to provide future ones as high a sustainable quality of life as it achieves. This may be called the maximum principle.

Societies may violate this principle either with respect to population characteristics or to quality of life. For example, if the population is too large or small to sustain the present quality of life, then it should be decreased or increased. Similarly, the quality of life may be too high to be sustainable no matter what the size of the population, in which case it ought to be lowered. Also, the quality of life may be lower than is sustainable. That would not violate an obligation to future generations but it would probably violate obligations to members of the present one. However, one need not worry about this case, because most people desire and seek at least as good a life for themselves as is morally permissible. Since population characteristics such as size and age structure are not inherently valuable, one evaluates them by their impact upon the quality of life. That is, one first determines the maximum sustainable quality of life and population characteristics are adjusted to provide for it, keeping in mind that the opportunity to have children is one factor in the quality of life.

An objection to such obligations to far distant generations is that one cannot know what they will desire or need.[16] One cannot provide for their quality of life because one does not know what to provide. This objection has more force with respect to providing for a particular type of goods and services than it does with respect to providing for a quality of life. Quality of life is a broad concept covering natural and social conditions for life's worthwhileness. One can be pretty sure future generations will need a source of food; education; social institutions preventing violent conflicts; physical mobility; and the abilities to see, hear, and communicate.[17] Moreover, since the present generation is obligated to provide as high a quality of life as it enjoys, in the face of ignorance about future needs and desires, it should leave open as much freedom and opportunity as possible to allow an equally high quality of life in the future. If one does not know whether one will need to do A, B, or C next week, one reasonably tries to leave open all options. Thus, if the present generation does not know whether future ones will need arable land, forests, etc., it should leave as much as possible for them.

The two principles of generational sacrifice may be applied to the issue of food aid. They require that the present generation make sacrifices if they are necessary to ensure future generations a minimum or equivalent sustainable quality of life. The difficulty in applying them is in determining what quality of life future generations would have. On Hardin's scenario, if food aid is given, there will be many more deaths by starvation in the future. What this result implies about the quality of life of that generation as a whole depends on the criteria used to determine the quality of life of a society. It is not possible here to develop an index for the quality of life of a society. Nonetheless, one may make a commonsense judgment that widespread famine, with concomitant social turmoil and human misery, would constitute a low quality of life. However, the alternative is current starvation and a lower present quality of life. On the maximum principle, the present generation ought to suffer current starvation to the extent that failure to do so would result in a lower quality of life in the future. In short, it should suffer a current famine if that will avoid a worse one, but it need not suffer any more starvation than that. On a minimum principle, one's obligation depends on what the minimum is and whether or not a future famine would reduce the quality of life below it. It would certainly not require permitting as much present starvation as the maximum principle, and it provides a weak basis for an argument against food aid.

A further difficulty is that it is not certain that the Hardin scenario of food aid would in fact materialize. First, it is not clear that food aid does contribute to larger future populations.[18] Second, if stringent programs of population control are implemented, then aid may eliminate an imminent famine and population control a later one. Whether population control can or will be effective in time is a matter about which experts disagree. One's willingness to provide aid depends upon one's attitude toward risk taking. If one is a gambler, then one will be willing to provide aid now in the hope that the future catastrophe will be avoided by policies taken in the meantime. If one is cautious, one will not take such chances and will not give aid now in order to be certain to avoid a later catastrophe.

OBLIGATIONS TO THE PRESENT GENERATION

In order to support food aid, one must also justify sacrifice within the present generation. Peter Singer has argued for a principle which would require major sacrifices by developed countries to avoid famines. He presents a stronger and a weaker version. The strong version is: "If it is in our power to prevent something bad from happening, without thereby sacrificing anything of comparable moral importance, we ought, morally, to do it." The weak version is: "If it is in our power to prevent something very bad from happening, without thereby sacrificing anything morally significant, we ought, morally, to do it."[19] While Singer thinks the strong version is justifiable, he primarily argues for the weak one because he believes the grounds for it are uncontroversial, yet adherence to it requires considerable change in the way of life in developed countries.

Singer takes the principle to be a rather obvious implication of the principle that one ought to relieve suffering. He makes a number of points about it, but only three points need be noted here.[20] First, the principle applies regardless of distance. Whether the people to whom something bad would happen are near or far is morally irrelevant. Second, it is irrelevant how many other people are in a similar position to avoid the bad occurrence. If many others have contributed or will contribute to avoiding the bad occurrence, then one need give proportionately less. Third, he claims that the Hardin type of objection does not affect the moral principle, only the form of aid which ought to be given.[21] Indeed, he is inclined to accept the Hardin argument. However, instead of supporting the withholding of all aid, he thinks it affects only the kind of aid which should be provided. It should be aid in population control and economic development rather than food relief.

Singer's second point, about the irrelevance of how many other people are in a similar position, appears to rest upon the same assumption as Nell's claim about stockholders and citizens being causally responsible for deaths. However, there is a difference. Singer's principle applies only to bad occurrences which it is in one's power to avoid. Hence, he assumes causal efficacy, whereas Nell is arguing for causal efficacy. No ordinary stockholder or citizen has the power alone to determine company or government policy. However, it is in his power to contribute varying amounts of his income or wealth to famine relief.

Singer's argument, unlike Nell's, focuses on the question of how much people are required to sacrifice to help those in need. That is, he treats the issue as one of the duty to rescue rather than the duty not to kill. Moreover, his stronger and weaker versions present the extremes of plausible moral positions. Singer's weak version essentially states that one has a duty to rescue provided one will not have to make any sacrifice. His strong version is that one has a duty to rescue so long as the risk of loss in the attempt is no greater than the risk of loss from not attempting it. When the risk of loss in attempting rescue is greater than that of not trying, it is unreasonable.

The issue is the extent of a *duty* to rescue, not merely its reasonableness. There are heroic rescues, those which a person is not obligated to make. Sometimes what people call heroic rescues are merely unreasonable attempts which turn out successfully. Morally, they are instances of lucky foolhardiness. If there are to be non-obligatory but reasonable heroic rescues, those involving supererogatory acts, then Singer's strong version must be rejected. No one is obligated to sacrifice himself below a minimum quality of life even to save the life of another.[22]

At this point, however, one must compare the benefits forgone and those gained. Starving people would have their lives saved by food aid.

Consequently, to provide such important benefits, it is justifiable that others forgo possible lesser benefits. As it is chiefly food which is at stake, and consumption above that which is nutritionally adequate has only secondary benefits in the way of more interesting and tasteful diet (and often secondary harms, such as increased heart disease due to high cholesterol intake), even losses above the nutritionally adequate level are relatively unimportant. Consequently, people in developed countries at least have an obligation to forgo any increases in their food diets in order to avoid famines in less developed ones.

In more practical terms, then, the conclusions are as follows: Developed countries ought to provide aid to less developed ones at least to the extent that it does not involve a sacrifice of their current quality of life. Even this principle implies that they should sacrifice possible increases in their quality of life, especially food, to provide a minimum one for those in less developed countries. Some sacrifice of current quality of life may even be justifiable if it is necessary to prevent starvation. However, material goods, especially fancy foods, are not necessarily the central elements of the quality of life. Non-material goods, such as love and friendship, may be more important. Consequently, a reduction in the number of consumer goods per capita in developed countries does not necessarily imply a lower quality of life. The obligation to provide such aid is qualified by the obligation to future generations—namely, that it not result in their having a quality of life below the minimum or lower than the present one, depending on whether a minimum or maximum principle is adopted.

The obligation to future generations also affects the distribution of aid. One might think that, morally, aid should be distributed equally to nations in proportion to the needs of their citizens. But this principle assumes either that it is possible to save everyone (all countries) or that it is better for all to suffer or die equally than for some to survive. Although some philosophers have done so in the past, no one now defends the claim that supplies should be equally distributed even if all will certainly die. Instead, the claim is that it is not certain that all countries cannot be saved—i.e., avoid famine both now and later.[23] If one accepts this claim, then whether one is willing to risk famine in all by an equal distribution rather than certain famine in some and avoidance of famine in others depends upon one's attitudes toward risk.

At this point, the lifeboat analogy is misleading. In lifeboats, there are determinate events which count as being saved—namely, being picked up by another ship or reaching habitable land. Moreover, there is usually a fair estimate of the maximum time required for one or the other event to occur. With respect to population and famine, these ideas do not apply. If population growth continues, then the perilous situation could continue indefinitely. There is no outside agency which will effect rescue. Moreover, the idea of being saved is unclear, because there is no sudden sufficiency of food and other goods. Everyone taking the risk equally is more plausible in the lifeboat, because it is always possible that an outside agency will intervene within a few hours and save all.

An example from a different type of situation may make the point clearer. Suppose three people are caught in a fire, and one knows there is time to pull only one of them out. Would one be morally required by the principle of equality to pull each person one-third of the way out rather than one person all the way out? If one knows no other help is forthcoming, it would be morally wrong not to save one person. Suppose, however, that there might be help coming very shortly, so that all three persons could be saved. In that case, it might be reasonable to pull each person one-third of the way out if that would provide a few more minutes during which additional help might arrive.

In the population situation, there is no other help coming. However, there are a few analogous factors. New breakthroughs in food production might occur, although dramatic ones are not very likely. There might also be dramatic reduction in population growth rates by decreases in fertility, although the evidence to date does not support optimism. Yet, some food production/supply projections do involve a catastrophe. The crucial area is southern Asia. If population continues to grow in accordance with present trends, its population in 2025 will equal the current world population. Moreover, even if one assumes that population control measures will have an impact before the end of this century, the projected food deficit by 2000 exceeds the "surplus" of North America and by 2025 the entire North American production.[24] Given these projections, reasonable people may differ as to whether a selective policy of food aid is justifiable. Perhaps some countries should not receive food aid, but only aid in population control to help reduce current and future starvation.

NOTES

I wish to thank Tom Shannon, Jim Smith, and Onora O'Neill for helpful comments on previous versions of this essay.

1. I discuss distribution issues more thoroughly in "Selective and Conditional Population Aid: Some Moral Issues," an Occasional Paper, Proj-

ect on Cultural Values and Population Policy (Hastings-on-Hudson, N.Y.: Institute of Society, Ethics and the Life Sciences, 1977).

2. "Lifeboat Earth," *Philosophy and Public Affairs,* Vol. 4 (Spring, 1975), p. 273.
3. Ibid., p. 274.
4. See *People* v. *Young,* 11 N.Y.2d 274, 229 N.Y.S.2d 1, 183 N.E.2d 319 (1962).
5. "Living on a Lifeboat," *BioScience,* Vol. 24 (October, 1974), pp. 561–68.
6. Daniel Callahan, "Doing Well by Doing Good: Garrett Hardin's Lifeboat Ethic," *Hastings Center Report,* Vol. 4 (December, 1974), p. 3.
7. "Living on a Lifeboat," pp. 565 (italics omitted), 567.
8. Garrett Hardin, "Carrying Capacity as an Ethical Concept," *Soundings,* Vol. 59 (Spring, 1976), p. 133.
9. "Doing Well by Doing Good," p. 4.
10. Joseph Fletcher, "Feeding the Hungry: An Ethical Appraisal," *Soundings,* Vol. 59 (Spring, 1976), p. 58.
11. See Joseph Fletcher, "Indications of Humanhood: A Tentative Profile of Man," *Hastings Center Report,* Vol. 2 (November, 1972), pp. 1–4; *idem,* "Four Indicators of Humanhood–The Enquiry Matures," *Hastings Center Report,* Vol. 4 (December, 1974), pp. 4–7.
12. H. Tristram Engelhardt, Jr., "Individuals and Communities, Present and Future: Towards a Morality in a Time of Famine," *Soundings,* Vol. 59 (Spring, 1976), p. 79.
13. Derek Parfit, "On Doing the Best for Our Children," in *Ethics and Population,* ed. by Michael D. Bayles (Cambridge, Mass.: Schenkman Publishing Co., 1976), pp. 101–102.
14. I elaborate the basic argument of this paragraph in "Harm to the Unconceived," *Philosophy and Public Affairs,* Vol. 5 (Spring, 1976), pp. 292–304.
15. John Rawls, *A Theory of Justice* (Cambridge, Mass.: Harvard University Press, Belknap Press, 1971), pp. 293–98; Henry Sidgwick, *The Methods of Ethics,* 7th ed. (London: Macmillan & Co., 1907), p. 381.
16. Martin P. Golding, "Obligations to Future Generations," *The Monist,* Vol. 56 (January, 1972), pp. 85–99; Robert L. Cunningham, "Ethics, Ecology, and the Rights of Future Generations," *Modern Age* (Summer, 1975), pp. 260–71.
17. See also Daniel Callahan, "What Obligations Do We Have to Future Generations?" *American Ecclesiastical Review,* Vol. 164 (April, 1971), pp. 273–75. See also B. M. Barry, "Justice Between Generations," in *Law, Morality and Society: Essays on Honour of H. L. A. Hart,* ed. by P. M. S. Hacker and J. Raz (Oxford: Clarendon Press, 1977), pp. 274–75.
18. Lester R. Brown, for example, has claimed that "good nutrition is the best contraceptive"; *In the Human Interest: A Strategy to Stabilize World Population* (New York: Norton & Co., 1974), p. 119.

19. Peter Singer, "Famine, Affluence, and Morality," *Philosophy and Public Affairs,* Vol. 1 (Spring, 1972), p. 231.
20. Ibid., pp. 231–34.
21. Ibid., p. 240. Hardin has privately remarked that he would support at least some forms of aid for population control.
22. For a good discussion of the duty to rescue, see John Kleinig, "Good Samaritanism," *Philosophy and Public Affairs,* Vol. 5 (Summer, 1976), pp. 382–407. Kleinig takes the failure to rescue as causing harm exacerbation, although unlike Nell he does not count the failure to give food aid as killing. By the analysis here, failure to rescue is a failure to benefit and does not involve killing. Nonetheless, people have a duty to provide others with major benefits, such as life, when to do so they need forgo only minor benefits for themselves.
23. Callahan, "Doing Well by Doing Good," p. 3.
24. Committee on World Food, Health, and Population, *Population and Food: Crucial Issues* (Washington, D.C.: National Academy of Sciences, 1975), p. 17.

18

WHITHER OUR ENERGY HERITAGE?

CLAUDE R. HOCOTT

I owe a duty.

Paul, Romans 1:14

There is no dearth of philosophical speculation regarding the obligation of any generation to the succeeding one or ones. It is almost axiomatic that the acceptance of benefactions imposes on the recipients an obligation toward others who might be similarly benefited. Further, when said recipients are responsible for the very presence of the future beneficiaries, the obligation seems to be doubly imposed. I believe strongly that the cultural progress of the human race is the result of ready acceptance of this principle by dedicated ancestors.

The present generation has no legal obligation to the next, but surely there is an ethical duty to leave the state of society equal to or better than it was when inherited. Consequently, no people should abandon precipitously the patterns of societal action that have proven successful in the past. For this reason, government policy makers must be urged to move cautiously in imposing untried new policies for tried and true ones. One is reminded of Cromwell's admonition to his ardent advisors who were anxious to invoke drastic changes in England's government policies: "I beg you, Gentlemen, in all humility, to have the grace to consider seriously the possibility that you may be wrong."

THE UNEARNED INCREMENT

The geometric growth of affluence since the beginning of time is one of the great wonders of man's existence. We are all recipients of an unearned increment of luxury, developed from the industry, thrift, ingenuity, and sacrifice of preceding generations. The growth has been such that each succeeding generation is more affluent than the preceding one. My own father liked to remind us that he grew up under circumstances that we, his children, would consider abject poverty. Similarly, his father before him, and so on into antiquity. Much of this process has been the result of the accomplishments of science and technology.

We need only remind ourselves of the trials undergone by the nation's great inventors, such as Whitney and McCormick, to recognize the blessings modern Americans enjoy because of their dedication, labor, and sacrifice to an ideal.[1] In 1800, before the invention of the cotton gin and the reaper, roughly 90 percent of the nation's populace lived on farms. It took nine farm families to provide food and fiber for themselves and for one other family. By 1900, this percentage of farm families had dropped to about 67. Because of the introduction of machines and the shift from manpower to horsepower, two farm families could feed themselves, their animals, and one other family. Even more dramatic changes occurred in the next fifty years, and the ratio was inverted. By 1950, only about one-third of the populace was engaged in farming. In addition, farm families were released from the dawn-to-dusk drudgery. Over this period, another power dimension came on the scene. Not only were farms mechanized, but fossil-fuel power had largely replaced horsepower. The petroleum revolution had arrived.

Today our nation is further blessed by the petroleum age and the attendant by-products of the petrochemical industry. With the availability of relatively inexpensive fertilizer, herbicides, fungicides, insecticides, rodenticides, antibiotics, etc., made from petroleum feedstocks, one farmer can now use petroleum-powered farm machinery and services to turn new, high-yield varieties of plants and animals into enough meat and wheat to feed nearly fifty people. So great has become our harvest that its storage has become an ethical problem and the export of farm surplus a political football. It must be noted, however, that this productivity is not the result of the energy of one farmer alone, but one farmer supplemented by many barrels of oil. It is estimated that ten units of fossil energy are now used to place one equivalent unit of food energy on our plates.[2]

This great productivity has been made possible by the dedicated research and development efforts of a relatively few of our nation's men and women. Each of these scientists and engineers possesses a high sense of ethical purpose and responsibility to make the lot of mankind healthier, happier, and less burdensome. The attendant affluence has become the common heritage of all the world in which petroleum-based energy and American technology can be made available.

OUR ENERGY HERITAGE:
Ascendency

The beginning of the petroleum industry in America is usually dated as 1859, with the Drake well in Pennsylvania. However, the nation entered the twentieth century with a coal-based energy economy. The modern petroleum industry is generally considered to have been born with Spindletop near Beaumont, Texas, in 1901. During the next decade, the automobile industry began its rise in importance, and World War I added impetus to gasoline-powered vehicles both on land and in the air. By 1920, the petroleum age was truly underway.

The history of the industry during this first quarter of this century was far from tranquil. Chaotic conditions existed which have been characterized as "boom and bust." A boom town quickly sprang up with each new wildcat discovery. The mad scramble for oil under the "rule of capture" led each lease holder to drill as many wells as practical, as quickly as possible. The result was a glut of oil on the market, with a rapid decrease in the price of oil. This situation was conducive to considerable waste of oil, sloppy procedures in field development, and an early depletion of the new field's flush productive capacity. As production declined, jobs disappeared, leases with scattered equipment and debris were abandoned, and boom towns became ghost towns. Further, as the oil production declined, demand drove prices upward again, and the economic incentive returned to trigger another wildcat exploratory program. As each one was successful, the cycle was repeated.

If the "boom and bust" cycle of oil field development was undesirable, the instability of the "feast or famine" supply picture was disastrous. A fortunate few reaped wealth from a successful venture, but the total industry's financial picture was not wholesome.[3] It has been estimated that during these early years, the total income from successful discoveries was far less than the ventured capital. Wildcatters were merely gambling at high risk with huge stakes. They have been truly called "the world's greatest gamblers."[4]

Early in the period of rapid growth in demand for petroleum, competent government people and concerned industry executives increasingly recognized the necessity of bringing some stability and continuity of supplies to the marketplace. This called for an orderly replacement of reserves commensurate with production rates. Replacement of reserves required successful operators to explore, discover, and develop new fields on a continuing basis. The successful operator, in turn, required an economic incentive to remain in the business.

CAPITAL GENERATION

Soon after the income tax statutes were implemented, Congress recognized that ample internal risk capital must be generated by the young, expanding petroleum industry in order for operators to conduct the level of exploration and development necessary to maintain the petroleum reserves at a level needed for a stable economy. Because the rate of discovery in the high-risk exploratory effort was low, even in those days, tax provisions were needed for operators to recover the cost of numerous unsuccessful exploratory ventures from the income earned on the relatively few productive fields. In 1926, after experience with several procedures that proved to be impractical, Congress adopted a tax provision allowing the deduction of 27.5 percent of gross revenue from production as a fair and equitable schedule for investment capital recovery. It is unfortunate that this provision was ever referred to as an "allowance" because depletion bears the same relation to an irreplaceable asset that depreciation does to replaceable property. Far from being a tax loophole, the 27.5 percent provision was a carefully considered and deliberately designed tax policy adopted by the Congress to permit oil operators to recover the capital requirements of a high-risk venture.[5] The wisdom of this decision was made clear by the subsequent unfolding history of the industry.

CONSERVATION

As has already been mentioned, many early practices of the petroleum industry were conducive to aboveground waste. This brought about legislative attempts to encourage the use of more efficient methods. As early as 1899, the State of Texas enacted a conservation statute to prevent wasteful oil field practices.[6] Amplifying President Hoover's famous definition, "Conservation is efficient use" to include "efficient production," the elimination of physical waste during production became the heart of conservation statutes, rules, and regulations. Most of these early efforts were devoted to the prevention of aboveground waste. A few oil industry pioneers, however, were already beginning to recognize the even more serious problem of underground waste, and they urged the oil-producing states to engage in concerted action for corrective measures.[7]

State autonomy and sovereignty posed major stumbling blocks to effective interstate cooperation. Many opponents of controls also insisted that individual state regulations to prevent waste were merely price-fixing devices. It became apparent by 1932, however, that some action was needed to reduce the inconsistencies and conflicts in rules and regulations that had been adopted by the various states. Consequently, the governors of Kansas, Oklahoma, New Mexico, and Texas appointed an Oil States Advisory Committee, which drafted a Uniform Legislative Act for Conservation and Interstate Compact. With strong support from Governors Marland of Oklahoma and Allred of Texas, plus an assist from President Roosevelt, the Interstate Compact to Conserve Oil and Gas was created in 1935 by act of the U.S. Congress. Its goal was to conserve and protect irreplaceable natural resources, prevent waste, and obtain the greatest ultimate recovery by and through state authority.[8]

The Compact reads, "The purpose of this compact is to conserve oil and gas by the prevention of waste thereof from any cause." The act further provides for each compacting state to enact reasonable legislation to accomplish that purpose. Of special significance is the specific provision of the act denying authorization of the compacting states to limit production or create monopolies for the purpose of price fixing. Also, the act specifies that no state shall become financially obligated to any other.

These wise laws brought results that exceeded the fondest expectations of their proponents. In the period from 1926 to 1956, crude oil reserves increased more than threefold, from 8.8 billion barrels to 30.4 billion barrels, and daily production rose from about 2 million to 7 million barrels.[9] Further, the nation had an excess, shut-in producing capacity of about 25 percent over demand. Texas was prorated to eight days of production per month under its market demand statute. As a direct result of this national posture, the Allies floated to victory in World War II on a sea of oil produced by U.S. technology and management. In addition, petroleum furnished practically all the toluene for Allied ammunition plus the raw material, research, and technology for a new synthetic rubber industry that to this day has kept us free of dependency on imported natural rubber.

A committee of the National Academy of Science reported in 1969:

The history of research and development for the mineral fluid industry has been a very gratifying one. Not only has this technology enabled a nation to have ample supplies of petroleum to fuel and lubricate a steadily expanding economy and a military operation through two world wars, but it has been able to do this through a long period of inflation in which petroleum products are among the few commodities which have not experienced spiraling prices. This impressive result has been accomplished by research and development financed almost wholly by private enterprise and disseminated freely and widely through professional and trade literature. It has provided the base for worldwide operations of the petroleum industry and as such is undoubtedly one of the United States' major technological exports.[10]

As Senator Eugene McCarthy stated recently:

If oil companies are to be faulted, it is for having provided the country with too much inexpensive oil and gasoline, in consequence of which wasteful, oversized automobiles were developed and wasteful fuel consumption practices followed both in industry and in consumer use.[11]

OUR ENERGY HERITAGE:
The Turn-around

It is almost inconceivable that we would willingly retreat from the successful policies that enabled our petroleum industry to provide the oil and gas so essential to the economic health of the nation and, indeed, the world. Nevertheless, such a deterioration began in 1954, when federal regulations implementing the Supreme Court's interpretation of the 1938 Natural Gas Act were written in such a way as to cover *producing* ventures as well as gas pipelines. The Federal Power Commission chose to ignore the congressional experience concerning the discovery value of oil and gas in 1918 and 1926. Further, the Commission adopted an unrealistic allocation of costs between oil and gas. In establishing a "fair market price," federal controls provided for a limited return on the book value of investments with no regard whatsoever for discovery value, much less for discovery value at the time of assessment. It was not, and still is not, recognized in government circles that anything less than replacement value (which is actually equally applicable during inflation or deflation) results inevitably in liquidation.[12]

This is exactly what has happened. Not only did liquidation proceed under these unrealistic controls, but it was accelerated by a tremendous increase in the national consumption of natural gas. The low, arbitrary control policies encouraged every energy user in the nation to utilize this premium fuel. Since there was no way for oil and coal to compete with the low unit heating value prices of natural gas, these industries shrank in the marketplace. However, the petroleum industry tried to remain competitive. In 1956, two years after the Supreme Court decision, a record 58,160 wells were drilled in an attempt to provide the nation with vital supplies of oil and gas. But the die was cast. Without the opportunity to generate risk capital to explore, discover, develop, and produce new oil and gas, the industry inescapably entered a prolonged drilling slump. The low point was reached in 1971, when only 27,300 wells were drilled.[13]

Fortunately for the nation, the industry had already become, to a limited extent, a multinational enterprise by 1954. Furthermore, under the international situation then existing, foreign oil could be discovered, developed, produced, and imported at prices that were competitive with cheap domestic gas. Consequently, large amounts of the oil industry's investment capital went abroad. Huge foreign reserves of oil and gas were developed which, in increasing amounts, supplied the oil needed to supplant our diminishing domestic reserves.

This condition prevailed until the oil embargo of 1973. History since that time is well known, but is confusing and misunderstood not only by the public but by the majority of public servants as well.

WHERE WE STAND

In the meantime, the federal government has blamed the industry for the nation's energy ills and has continued to strike out with punitive and counterproductive actions that further reduce the likelihood of any early remedy. Federal agencies loudly proclaim the malaise and yet continue to make decisions that ensure not only its continuance but a worsening of the situation.

Our nation has moved and continues to move deliberately away from oil industry policies that have proven enormously successful in the past. By this action, domestic supplies of oil and gas have been placed prematurely in jeopardy. The industry has lost twenty years of momentum in looking for new deposits and bringing them to production. Even worse, the development of knowledge regarding the location and methods of exploration for scarcer new deposits has been impeded. Also, the application of new technology for the future oil and gas resources has been retarded.

In addition to hampering the petroleum industry, federal energy policies have seriously discouraged the development of alternate domestic energy resources so vital to both the protection of consumer interests and to our national security.[14] As an example, there was no possibility that the more abundant coal reserves of the nation could be developed in the face of a deliberate policy of marketing premium natural gas at a regulated price, far below that at which coal could be brought to the market. Consequently, there has been inadequate research on methods to improve coal as an alternate fuel either for clean, direct burning or as a source of synthetic liquid and gas. The same conditions existed for vast oil shale deposits and various other potential sources of energy.

WHAT OF THE FUTURE?

At this time, the energy future must be gauged by expectations of national energy policy. It is not bright. Indications point to still further interference with the fundamental economic forces that control the energy picture. There are deliberate efforts to add additional barriers and disincentives to increasing the supplies of domestic oil and gas or to the direct burning of coal to fuel our national economy. These are the only energy sources that have the capability of filling the domestic energy gap in the near-term, i.e., the period 1985–2000. This gap must be filled in order that we may gain time to develop alternatives for this and succeeding generations. As a matter of fact, the current situation has the potential for broad-scale human tragedy.

It seems certain that, in the energy heritage this generation is passing to the next, we have foreclosed the opportunity that they, too, might have an abundant supply of energy at costs that would afford a prosperous economy. In other words, we have, in all probability, "killed the goose that laid the golden egg." As we move deliberately away from a pattern of energy enterprise which was so successful in the past, I wonder if we should not be more concerned about the ethics of foreclosing important options for our posterity!

AN ALTERNATE ENERGY FUTURE

Observers of the national scene frequently comment that we have *no* energy policy.[15] I hope that our government leaders have not maligned the energy industry to such an extent that recovery is politically impossible. The petroleum industry must be provided with effective incentives in order to recover the momentum of earlier days.

Most energy experts agree that oil, gas, and/or coal, by direct burning, afford the only possible energy resources with the potential of meeting the energy requirements of a healthy national economy for the remainder of the twentieth century. They agree that this time period is the minimum recovery needed to develop viable alternative energy sources. Still further, considering environmental and other societal barriers, it is probably equally true that the economy of the country cannot afford the capital requirements to shift rapidly much of our petroleum-based facilities to coal utilization. In the face of all these factors, it seems inevitable that the nation will be forced to continue heavy reliance on oil and gas for the next decade or more. How, then, should the nation respond to this conclusion?

First, our people must face the fact that the days of really inexpensive energy are gone, perhaps forever. Next, we must realize that the nation will have to rely strongly on oil and gas which must either be produced domestically or be imported. Surely then, it would be widely popular to adopt an energy policy that provides ample incentive for petroleum companies to proceed with a great sense of urgency to explore for, discover, develop, and produce some of this nation's remaining potential hydrocarbon resources. There was nothing in the discovery rate in 1956 to lead to the conclusion that the end of attractive petroleum resources was in sight. In fact, a recent publication makes a very persuasive argument that if the drilling rate of 1956 had prevailed until 1973, the oil embargo would have been impossible.[16] There is, in fact, the distinct probability that if industry drilling programs had continued the growth rate of 1946–1956, the nation would still have petroleum in storage for emergencies in the form of excess producing capacity. During this decade, the total wells drilled by industry increased from 30,230 to 58,160, and remaining crude oil reserves increased by about 9 billion barrels, while total production exceeded 19 billion barrels. Even during the period of serious reduction in drilling from 1956 to 1971, 39.8 billion barrels of crude oil were added to reserves while 41.8 billion barrels were produced, leading to a reduction of net reserves of only about 2 billion barrels.[17]

An interesting sidelight of the above scenario is that, in addition to maintaining a healthy discovery rate, the industry would have continued to progress along the learning curve as to the location of petroleum resources and the development of exploration technology required to find it. Also, the industry could have generated the capital and the motivation to conduct more extensive research on alternate energy sources. In this event, the situation for the next generation would have been immeasurably improved by a rational approach that would also have been best for the present. This seems paradoxical, does it not? But that's the way with ethics.

NOTES

1. D. J. Breeden, ed., *Those Inventive Americans* (Washington, D.C.: National Geographic Society, 1971), pp. 64ff.
2. C. E. Steinhart and J. S. Steinhart, *Energy: Sources, Use, and Role in Human Affairs* (Belmont, Calif.: Duxbury Press, 1974), p. 90.
3. J. A. Clark, *Three Stars for the Colonel* (New York: Random House, 1954), p. 165.
4. R. S. Knowles, *The Greatest Gamblers* (New York: McGraw-Hill Book Co., 1959).
5. H. H. Baker, *Tax Policy and Petroleum Supplies* (American Petroleum Institute, Division of Production, 1950), p. 43; R. J. Gonzales, *Taxation of Petroleum* (Fifth Annual Colorado Petroleum Council Meeting, June, 1964); "Percentage Depletion—A History," Special Task Force Report, Mid-Continent Oil and Gas Association (March, 1968), p. 72.
6. Clark, *op. cit.*
7. W. E. Pratt, *The Basis of Proration in Texas*, American Association of Petroleum Geologists (September, 1939), p. 1314.
8. J. J. Matthews, *Life and Death of an Oilman* (Norman, Okla.: University of Oklahoma Press, 1951), pp. 216ff.; R. J. Sullivan, ed., *Conservation of Oil and Gas*, Section of Mineral and Natural Resources Law (American Bar Association, 1958), p. 279.
9. *Petroleum Facts and Figures* (New York: American Petroleum Institute, 1967).
10. "Mineral Science and Technology," Report of the National Academy of Science (Washington, D.C.: Government Printing Office, 1969), p. 39.
11. Senator Eugene McCarthy, quoted from "International Oil Scouts Association—Official Publication," August, 1976.
12. M. A. Adelman, "The Supply and Price of Natural Gas," *Supplement to Journal of Industrial Economics* (Oxford: Basil Blackwell, Ltd., 1962), pp. 8ff.; R. J. Gonzalez, "Impact of Government Policies on Petroleum Supplies," *Exploration and Economics of Petroleum Industries*, Vol. 10 (New York: Matthew Bender, Inc., 1972), pp. 209ff.
13. *Petroleum Facts and Figures.*
14. Gonzales, *op. cit.*
15. "Mineral Science and Technology."
16. Gonzales, *op. cit.*
17. *Petroleum Facts and Figures.*

THE CONTRIBUTORS

MICHAEL D. BAYLES is Professor of Philosophy at the University of Kentucky. His publications include *Contemporary Utilitarianism, Ethics and Population, Medical Treatment of the Dying: Moral Issues,* and *Principles of Legislation.* Professor Bayles serves on the Advisory Board of the *Southern Journal of Philosophy.*

EDMUND BEARD is Assistant Professor of Politics at the University of Massachusetts–Boston. His publications include *Congressional Ethics: The View from the House, The Basics of American Politics,* and *Developing the ICBM: A Study in Bureaucratic Politics.*

WILLIAM BLACKSTONE (1931–1977) was Research Professor of Philosophy and Chairman of the Division of Social Sciences at the University of Georgia. He authored or edited numerous scholarly articles plus several books, including *Ethics and Education, The Concept of Equality,* and *Philosophy and Environmental Crisis.* Dr. Blackstone was a past president of the American Society for Value Inquiry.

ABRAHAM J. BRILOFF is Emanuel Saxe Distinguished Professor of Accountancy at Baruch College, City University of New York. Dr. Briloff is the author of numerous scholarly articles plus three books: *The Effectiveness of Accounting Communication, Unaccountable Accounting,* and *More Debits than Credits.* He has received the Townsend Harris and the 125th Anniversary Distinguished Alumnus Medals from City College.

JOHN F. BURTON, JR., is Professor of Industrial Relations and Public Policy at the Graduate School of Business, University of Chicago. He has been a senior staff economist for the Council of Economic Advisers and Chairman of the National Commission on State Workmen's Compensation Laws. Dr. Burton is the author of numerous scholarly articles and is co-editor of *Readings in Labor Market Analysis.*

RICHARD DE GEORGE is University Professor of Philosophy at the University of Kansas. He is the author or editor of ten books, including *Ethics and Society, The New Marxism,* and *Soviet Ethics and Morality.*

VICTOR GOTBAUM is Executive Director, District Council 37, of the American Federation of State, County and Municipal Employees, AFL-CIO. This union, with 110,000 members, is the largest in New York and the largest union of municipal employees in the country. Mr. Gotbaum is also vice president of the New York City Central Labor Council, AFL-CIO.

ROBERT W. GREEN is an associate in the Washington, D.C., law firm of Arent, Fox, Kintner, Plotkin & Kahn. This firm represents over forty-five national trade associations in legal matters. Mr. Green cooperated with Earl W. Kintner in preparing the article "Legal Limitations and Possibilities for Self-Enforcement of Codes of Ethics," which appears in *The Ethical Basis of Economic Freedom.*

MICHAEL HARRINGTON, Chair of the Democratic Socialist Organizing Committee, is the author of *The Other America, Socialism, The Twilight of Capitalism,* and other books. He serves on the editorial board of *Dissent* and teaches at Queens College, City University of New York. Mr. Harrington was Chairman of the Socialist Party of the United States from 1968 to 1972.

CLAUDE R. HOCUTT is Professor of Petroleum Engineering at the University of Texas–Austin. He has served as Executive Vice President of Esso Production Research Company, Houston. Professor Hocott has written and lectured widely on the subjects of resource exploration and management. He is listed in *American Men and Women of Science.*

JOHN HOSPERS is Professor in the University of Southern California School of Philosophy. He is the editor of *The Personalist* and the author of numerous books, including *An Introduction to Philosophical Analysis, Human Conduct,* and *Libertarianism: A Political Philosophy Whose Time Has Come.* Professor Hospers was the Libertarian Party's first candidate for President of the United States in 1972.

WILLIAM J. KILBERG served as Solicitor, U.S. Department of Labor, from 1973 until 1977. He was formerly Special Assistant to the Secretary of Labor, Counsel for the Federal Mediation and Conciliation Service, and a White House Fellow. Mr. Kilberg is the author of numerous articles on labor law and policy and a contributing author to *The Lessons of Victory.* He is presently a partner in the Washington, D.C. office of the New York City law firm, Breed, Abbott and Morgan.

EARL W. KINTNER is a senior partner in the Washington, D.C. law firm of Arent, Fox, Kintner, Plotkin & Kahn. He is a former Chairman and General Counsel of the Federal Trade Commission. Mr. Kintner is the author of six widely acclaimed primers on the federal antitrust laws and the definitive multivolume legislation history of these laws: *Federal Antitrust Laws and Related Statutes* (1978).

IRVING KRISTOL is Henry R. Luce Professor of Urban Values at New York University, co-editor of *The Public Interest,* and a member of the *Wall Street Journal's* "Board of Contributors." He is the author of numerous articles and is co-editor of four books. A collection of his essays, entitled *On the Democratic Idea in America,* appeared in 1972, and in 1978 he published *Two Cheers for Capitalism.*

JOHN LACHS is Professor of Philosophy at Vanderbilt University. He is author of *Animal Faith and Spiritual Life* and *The Ties of Time,* co-editor of *Physical Order and Moral Liberty,* and contributor to *Challenges in American Culture.* He is past president of the Society for the Advancement of American Philosophy.

BURTON M. LEISER is Professor and Chairman of the Department of Philosophy at Drake University. His numerous publications include *Liberty, Justice and Morals* and *Custom, Law and Morality.*

JOSEPH W. McGUIRE is Professor of Administration at the University of California, Irvine. He is a frequent contributor to scholarly journals in the areas of economics, management, and management responsibility. In addition, Professor McGuire has authored or edited fifteen books, including *Contemporary Management, Business and Society,* and *Theories of Business Behavior.* Dr. McGuire was formerly vice-president for planning for the University of California system.

PHILLIP J. NELSON is Professor of Economics at the State University of New York at Binghamton. He has published numerous scholarly articles on the economics of information and advertising. Professor Nelson is currently a Fellow at the Hoover Institution on War, Revolution and Peace.

JOSEPH A. PICHLER is Dean and Professor, University of Kansas School of Business. He is a frequent contributor to scholarly journals in the areas of labor relations and human resources. Dr. Pichler is co-author of *Inequality: The Rich and Poor in America.*

INDEX